Ben Stacy Jerrik (Ed.)

Tennessee State Route 840

Ben Stacy Jerrik (Ed.)

Tennessee State Route 840

State highway, Nashville, Tennessee, Lamar Alexander, Tennessee Department of Transportation, Interstate 40

Part Press

Imprint

Publisher:
Part Press is a trademark of
International Book Market Service Ltd., 17 Rue Meldrum, Beau Bassin, 1713-01 Mauritius
Email: info@bookmarketservice.com
Website: www.bookmarketservice.com

Published in 2013

Printed in: U.S.A., U.K., Germany. This book was not produced in Mauritius.

ISBN: 978-613-9-37645-2

Contents

References

Red_Buttons

Red Buttons	
 Buttons in 1978	
Born	Aaron ChwattFebruary 5, 1919New York City, New York, U.S.
Died	July 13, 2006 (aged 87)Century City, Los Angeles, California, U.S.
Occupation	Actor, comedian
Years active	1935–2006
Spouse(s)	Roxanne Arlen (m. 1947 – 1949) Helayne McNorton (m. 1949 – 1963) Alicia Prats (m. 1964 – 2001)

Red Buttons (born **Aaron Chwatt**, February 5, 1919 – July 13, 2006) was an Academy Award-winning American comedian and actor.

Early life

Red Buttons was born **Aaron Chwatt**[1] on February 5, 1919 in New York City to Jewish immigrants Sophie (née Baker) and Michael Chwatt.[2] [3] At sixteen years old, Chwatt got a job as an entertaining bellhop at Ryan's Tavern in City Island, Bronx. The combination of his red hair and the large, shiny buttons on the bellhop uniforms inspired orchestra leader Charles "Dinty" Moore to call him "Red Buttons," the name under which he would later perform.

Later that same summer, Buttons worked on the Borscht Belt;[1] his straight man was Robert Alda. In 1939, Buttons started working for Minsky's Burlesque; in 1941, José Ferrer chose Buttons to appear in a Broadway show *The Admiral Had a Wife*. The show was a farce set in Pearl Harbor, and it was due to open on December 8, 1941. It never did, as it was deemed inappropriate after the Japanese attack. In later years, Buttons would joke that the Japanese only attacked Pearl Harbor to keep him off Broadway.

Career

In September 1942, Buttons made his Broadway debut in *Vickie* with Ferrer and Uta Hagen. Later that year, he appeared in the Minsky's show *Wine, Women and Song;* this was the last Burlesque show in New York City history, as the Mayor La Guardia administration closed it down. Buttons was on stage when the show was raided.

Drafted into the United States Army Air Forces, Buttons in 1943 appeared in the Army Air Forces' Broadway show *Winged Victory*, along with several future stars, including Mario Lanza, John Forsythe, Karl Malden and Lee J. Cobb. A year later he appeared in Darryl F. Zanuck's movie version of *Winged Victory*, directed by George Cukor. Buttons also entertained troops in the European Theater in the same unit as Mickey Rooney.

After the war, Buttons continued to do Broadway shows. He also performed at Broadway movie houses with the Big Bands. In 1952, Buttons received his own variety series on television, *The Red Buttons Show*," which ran for three years, and achieved high levels of success. In 1953, he recorded and had a two-sided hit with *Strange Things Are Happening/The Ho Ho Song*, with both sides/songs essentially being the same.

His role in *Sayonara* was a dramatic departure from his previous work. In that film, co-starring with Marlon Brando, he played Joe Kelly, an American airman stationed in Kobe, Japan during the Korean War, who falls in love with Katsumi, a Japanese woman (played by Miyoshi Umeki), but is barred from marrying her by military rules intended to reassure the local populace that the U.S. presence is temporary. His moving portrayal of Kelly's calm resolve not to abandon the relationship, and the touching reassurance of Katsumi, impressed audiences and critics alike; both he and Umeki won supporting actor and actress Academy Awards for the film.

After his Oscar-winning role, Buttons performed in numerous feature films, including the Africa adventure *Hatari!* with John Wayne, the war epic *The Longest Day*, the biopic *Harlow*, the disaster film *The Poseidon Adventure*, the dance-marathon drama *They Shoot Horses, Don't They?*, the family comedy *Pete's Dragon*, the disaster film *When Time Ran Out* with Paul Newman and the age-reversal comedy *18 Again!* with George Burns.

Buttons as Henry Phyfe.

In 1966, Buttons again starred in his own TV series, a spy spoof called *The Double Life of Henry Phyfe*, which ran for one season. Buttons also made many memorable guest television appearances on programs including *The Eleventh Hour*, *Little House on the Prairie*, *It's Garry Shandling's Show*, *ER* and *Roseanne*. His last regular role was as a homeless man on CBS' *Knots Landing*.

He became a nationally recognizable comedian, and his "Never Got A Dinner" routine was a standard of The Dean Martin Celebrity Roasts for many years. He was number 71 on Comedy Central's list of the *100 Greatest Stand-Ups of All Time*.

Buttons received a star on the Hollywood Walk of Fame for television, his star being located at 1651 Vine Street.

Personal life

Buttons married actress Roxanne Arlen in 1947, but it soon ended in divorce. He married Helayne McNorton on December 8, 1949, and they divorced 1963. His last marriage was to Alicia Pratt, and lasted from January 27, 1964 until her death in March 2001. Buttons had two children, Amy Buttons and Adam Buttons. He was the advertising spokesman for the Century Village, Florida retirement community.

Buttons was an early member of the Synagogue for the Performing Arts, and at the time, Rabbi Jerome Cutler was the Rabbi.[4]

Death

Red Buttons died of complications from high blood pressure in 2006 at his home in Century City, Los Angeles. He had long been ill and was with family members when he died.[1]

Filmography

Year	Title	Role	Notes
1947	*Winged Victory*	Whitey/Andrews Sister	as Cpl. Red Buttons
1947	*13 Rue Madeleine*	Second Jump Master	uncredited
1951	*Footlight Varieties*	Himself	
1957	*Sayonara*	Airman Joe Kelly	Academy Award for Best Supporting Actor Golden Globe Award for Best Supporting Actor - Motion Picture Nominated—BAFTA Award for most promising newcomer
1958	*Imitation General*	Cpl. Chan Derby	
1959	*The Big Circus*	Randy Sherman	
1961	*One, Two, Three*	MP sergeant	uncredited
1962	*Hatari!*	Pockets	
1962	*Five Weeks in a Balloon*	Donald O'Shay	
1962	*The Longest Day*	Pvt. John Steele	
1962	*Gay Purr-ee*	Robespierre	voice
1963	*A Ticklish Affair*	Uncle Cy	
1964	*Your Cheatin' Heart*	Shorty Younger	
1965	*Up from the Beach*	Pfc. Harry Devine	
1965	*Harlow*	Arthur Landau	Nominated—Golden Globe Award for Best Supporting Actor - Motion Picture
1966	*Stagecoach*	Peacock	
1969	*The Moviemakers*	Himself	short subject
1969	*They Shoot Horses, Don't They?*	Sailor	Nominated—Golden Globe Award for Best Supporting Actor - Motion Picture
1971	*Who Killed Mary What's 'Er Name?*	Mickey	
1972	*The Poseidon Adventure*	James Martin	
1976	*Gable and Lombard*	Ivan Cooper	
1977	*Viva Knievel!*	Ben Andrews	
1977	*Pete's Dragon*	Hoagy	Nominated—Saturn Award for Best Supporting Actor
1978	*Movie Movie*	Peanuts/Jinks Murphy	
1978	*The Users*	Warren Ambrose	TV movie
1979	*C.H.O.M.P.S*	Bracken	
1979	*Rudolph and Frosty's Christmas in July*	Voice of Milton	
1980	*When Time Ran Out*	Francis Fendly	

1985	*Alice in Wonderland*	The White Rabbit	
1988	*18 Again!*	Charlie	
1990	*The Ambulance*	Elias Zacharai	
1994	*It Could Happen to You*	Walter Zakuto	
1999	*The Story of Us*	Arnie Jordan	
2001	*Odessa or Bust*	The Old Man	short subject
2004	*Goodnight, We Love You*		documentary
2005	*Sid Bernstein Presents...*	Himself	documentary

References

[1] *Comedian Red Buttons dies at 87* (http://news.bbc.co.uk/1/hi/world/americas/5178496.stm). BBC News. 14 July 2006.

[2] "Motion Pictures". *Encyclopaedia Judaica*. Keter Publishing House. 1971-1972.

[3] http://www.filmreference.com/film/55/Red-Buttons.html

[4] http://www.forward.com/articles/8150/ *The Forward*

External links

- Interview with Red Buttons Television Writer, August 2012 (http://classicshowbiz.blogspot.com/2012/08/an-interview-with-ben-starr-part-two.html)
- Red Buttons (http://www.imdb.com/name/nm999/) at the Internet Movie Database
- Red Buttons (http://tcmdb.com/participant/participant.jsp?participantId=26337) at the TCM Movie Database
- Red Buttons (http://www.ibdb.com/person.asp?ID=33909) at the Internet Broadway Database
- Red Buttons (http://www.allrovi.com/name/p9969) at AllRovi
- Actor Red Buttons dead at 87 (http://www.cbc.ca/arts/story/2006/07/13/red-buttons-obit.html)
- Red Buttons (http://www.findagrave.com/cgi-bin/fg.cgi?page=gr&GRid=14915440) at *Find a Grave*
- Interview (http://www.youtube.com/watch?v=xJoPPDuTuF4) by Leon Charney on The Leon Charney Report

Academy_Award

WARNING: Article could not be rendered - ouputting plain text.

Potential causes of the problem are: (a) a bug in the pdf-writer software (b) problematic Mediawiki markup (c) table is too wide

Academy Award 85th Academy AwardsAn Academy Award statuetteAwarded for Excellence in cinematic achievementsPresented byAcademy of Motion Picture Arts and SciencesCountry United StatesFirst awarded 1929Official website www.oscars.orgThe Academy Awards, informally known as The Oscars, are a set of awards given annually for excellence of cinematic achievements. The Oscar statuette is officially named the Academy Award of Merit and is one of nine types of Academy Awards. Organized and overseen by the Academy of Motion Picture Arts and Sciences (AMPAS), "About the Academy Awards". Academy of Motion Picture Arts and Sciences. Archived from the original on April 7, 2007. . Retrieved April 13, 2007. the awards are given each year at a formal ceremony. The AMPAS was originally conceived by Metro-Goldwyn-Mayer studio executive Louis B. Mayer as a professional honorary organization to help improve the film industry's image and help mediate labor disputes. The awards itself was later initiated by the Academy as an award "of merit for distinctive achievement" in the industry.Essex, Andrew (May 14, 1999). "The Birth of Oscar". Entertainment Weekly. . Retrieved March 2, 2011.The awards were first given in 1929 at a ceremony created for the awards, at the Roosevelt Hotel (Hollywood)Hotel Roosevelt in Hollywood. Over the years that the award has been given, the categories presented have changed; currently Oscars are given in more than a dozen categories, and include films of various types. As one of the most prominent award ceremonies in the world, the Academy Awards ceremony is televised live in more than 100 countries annually. It is also the oldest award ceremony in the media; its equivalents, the Grammy Awards (for music), the Emmy Awards (for television), and the Tony Awards (for theater), are modeled after the Academy Awards. The 84th Academy Awards were held at the Dolby Theatre on February 26, 2012; the 85th Academy Awards are scheduled to be held on February 24, 2013, in the same venue. History Gary Cooper and Joan Fontaine holding their Oscars at the Academy Awards, 1942 The first awards were presented on May 16, 1929, at a private brunch at the Hollywood Roosevelt Hotel with an audience of about 270 people. The post Academy Awards party was held at the Mayfair Hotel. "History of the Academy Awards". Academy of Motion Picture Arts and Sciences. . The cost of guest tickets for that night's ceremony was $5. Fifteen statuettes were awarded, honoring artists, directors and other personalities of the filmmaking industry of the time for their works during the 1927–1928 period. Winners had been announced three months earlier; however, that was changed in the second ceremony of the Academy Awards in 1930. Since then and during the first decade, the results were given to newspapers for publication at 11 pm on the night of the awards. This method was used until the Los Angeles Times announced the winners before the ceremony began; as a result, the Academy has since 1941 used a sealed envelope to reveal the name of the winners.For the first six ceremonies, the eligibility period spanned two calendar years. For example, the 2nd Academy Awards presented on April 3, 1930, recognized films that were released between August 1, 1928 and July 31, 1929. Starting with the 7th Academy Awards, held in 1935, the period of eligibility became the full previous calendar year from January 1 to December 31. The first Best Actor awarded was Emil Jannings, for his performances in The Last Command (1928 film)The Last Command and The Way of All Flesh (film)The Way of All Flesh. He had to return to Europe before the ceremony, so the Academy agreed to give him the prize earlier; this made him the first Academy Award winner in history. The honored professionals were awarded for all the work done in a certain category for the qualifying period; for example, Jannings received the award for two movies in which he starred during that period. Since the fourth ceremony, the system changed, and professionals were honored for a specific performance in a single film. As of the 83rd Academy Awards ceremony held in 2011, a total of 2,809 Oscars have been given for 1,853 awards. "A Brief History of the Oscar". Academy of Motion Picture Arts and Sciences. . Retrieved January 23, 2012. A total of 302 actors have won Oscars in competitive acting categories or have been awarded Honorary or Juvenile Awards.The 1939 film Beau Geste (1939 film)Beau Geste is the only movie that features as many as four Academy Award winners for Best Actor in a Leading Role (Gary Cooper, Ray Milland, Susan Hayward, Broderick Crawford) prior to any of the actors receiving the Best Actor Award.At the 29th ceremony, held on March 27, 1957, the Best Foreign Language Film category was introduced. Until then, foreign-language films were honored with the Special Achievement Award. Oscar Statuette Design Although there are eight other types of annual awards presented by the Academy (the Irving G. Thalberg Memorial Award, the Jean Hersholt Humanitarian Award, the Gordon E. Sawyer Award, the Academy Scientific and Technical Award, the Academy Award for Technical Achievement, the John A. Bonner Medal of Commendation, and the Student Academy Award) plus two awards that are not presented annually (the Special Achievement Award in the form of an Oscar statuette and the Honorary Award that may or may not be in the form of an Oscar statuette), the best known one is the Academy Award of Merit more popularly known as the Oscar statuette. Made of gold-plated britannia metalbritannium on a black metal base, it is 13.5 in (34 cm) tall, weighs 8.5 lb (3.85 kg) and depicts a knight rendered in Art Deco style holding a crusadecrusader's sword standing on a reel of film with five spokes. The five spokes each represent the original branches of the Academy: Actors, Writers, Directors, Producers, and Technicians. "Oscar Statuette: Legacy". Academy of Motion Picture Arts and Sciences. . Retrieved April 13, 2007.In 1928, MGM's art director Cedric Gibbons, one of the original Academy members, supervised the design of the award trophy by printing the design on a scroll. "Academy to Commemorate Oscar Designer Cedric Gibbons" (Press release). Academy of Motion Picture Arts and Sciences. May 3, 2007. . Retrieved April 13, 2007. In need of a model for his statuette, Gibbons was introduced by his future wife Dolores del Río to Mexican film director and actor Emilio FernándezEmilio "El Indio" Fernández. Reluctant at first, Fernández was finally convinced to pose nude to create what today is known as the "Oscar". Then, sculptor George Stanley (sculptor)George Stanley (who also did the Muse Fountain "Muse Fountain". . at the Hollywood Bowl) sculpted Gibbons's design in clay and Sachin Smith cast the statuette in 92.5 percent tin and 7.5 percent copper and then gold-plated it. The only addition to the Oscar since it was created is a minor streamlining of the base. The original Oscar mold was cast in 1928 at the C.W. Shumway & Sons Foundry in Batavia, Illinois, which also contributed to casting the molds for the Vince Lombardi Trophy and Emmy Awards statuettes. Since 1983, "Eladio Gonzalez sands and buffs Oscar #3453". Boston Globe. February 20, 2009. Archived from the original on February 23, 2009. . Retrieved February 21, 2009. approximately 50 Oscars are made each year in Chicago by Illinois manufacturer R.S. Owens & Company.Babwin, Don (January 27, 2009). "Oscar 3453 is 'born' in Chicago factory". Associated Press. Archived from the original on January 27, 2009. .In support of the American effort in World War II, the statuettes were made of plaster and were traded in for gold ones after the war had ended. "Oscar Statuette: Manufacturing, Shipping and Repairs". Academy of Motion Picture Arts and Sciences. Archived from the original on September 27, 2007. . Retrieved April 13, 2007.Naming The root of the name Oscar is contested. One biography of Bette Davis claims that she named the Oscar after her first husband, band leader Harmon Oscar Nelson; "Bette Davis biography". The Internet Movie Database. Archived from the original on March 8, 2007. . Retrieved April 13, 2007. one of the earliest mentions in print of the term Oscar dates back to a Time (magazine)Time magazine article about the 1934 6th Academy Awards. "Cinema: Oscars". Time. March 26, 1934. . Walt Disney is also quoted as thanking the Academy for his Oscar as early as 1932. "Oscar®-Winning Walt". Disney.Go.com. . Retrieved February 25, 2012. Another claimed origin is that the Academy's Executive Secretary, Margaret Herrick, first saw the award in 1931 and made reference to the statuette's reminding her of her "Uncle Oscar" (a nickname for her cousin Oscar Pierce)."Oscar" in The Oxford English Dictionary, June 2008 Draft Revision. Columnist Sidney Skolsky was present during Herrick's naming and seized the name in his byline, "Employees have affectionately dubbed their famous statuette 'Oscar'".Levy, Emanuel (2003) All About Oscar: The History and Politics of the Academy Awards Continuum, New York. ISBN 0-8264-1452-4 The trophy was officially dubbed the "Oscar" in 1939 by the Academy of Motion Pictures Arts and Sciences.

"OSCAR.com – 80th Annual Academy Awards – Oscar Statuette". Academy of Motion Picture Arts and Sciences. . Another legend reports that the Norwegian-American Eleanor Lilleberg, executive secretary to Louis B. Mayer, saw the first statuette and exclaimed, "It looks like King Oscar II!". In Honor of King Oscar II of Norway by Jørn K. Baltzersen, LewRockwell.comOwnership of Oscar statuettes Since 1950, the statuettes have been legally encumbered by the requirement that neither winners nor their heirs may sell the statuettes without first offering to sell them back to the Academy for US$1. If a winner refuses to agree to this stipulation, then the Academy keeps the statuette. Academy Awards not protected by this agreement have been sold in public auctions and private deals for six-figure sums.(Levy 2003, pg 28) In December 2011, Orson Welles' 1941 Oscar for Citizen Kane was put up for auction, after his heirs won a 2004 court decision that Welles did not sign any agreement to return the statue to the Academy.Duke, Alan (December 12, 2011). "Orson Welles' 'Citizen Kane' Oscar for sale". CNN. . Retrieved = December 12, 2011.While the Oscar is under the ownership of the recipient, it is essentially not on the open market.Lacey Rose (February 28, 2005). "Psst! Wanna Buy An Oscar?". Forbes. . Retrieved April 13, 2007. The case of Mike ToddMichael Todd's grandson trying to sell Todd's Oscar statuette illustrates that there are some who do not agree with this idea. When Todd's grandson attempted to sell Todd's Oscar statuette to a movie prop collector, the Academy won the legal battle by getting a permanent injunction. Although Oscar sales transactions have been successful, some buyers have subsequently returned the statuettes to the Academy, which keeps them in its treasury.(Levy 2003, pg 29)Nomination Since 2004, Academy Award nomination results have been announced to the public in late January. Prior to that, the results were announced in early February. Voters The Academy of Motion Picture Arts and Sciences (AMPAS), a professional honorary organization, maintains a voting membership of 5,783 as of 2012.Sandy Cohen (January 30, 2008). "Academy Sets Oscars Contingency Plan". AOL News. . Retrieved March 19, 2008.Academy membership is divided into different branches, with each representing a different discipline in film production. Actors constitute the largest voting bloc, numbering 1,311 members (22 percent) of the Academy's composition. Votes have been certified by the auditing firm PricewaterhouseCoopers (and its predecessor Price Waterhouse) for the past 73 annual awards ceremonies.Jackie Finlay (March 3, 2006). "The men who are counting on Oscar". BBC News. Archived from the original on March 19, 2007. . Retrieved April 13, 2007.All AMPAS members must be invited to join by the Board of Governors, on behalf of Academy Branch Executive Committees. Membership eligibility may be achieved by a competitive nomination or a member may submit a name based on other significant contribution to the field of motion pictures. New membership proposals are considered annually. The Academy does not publicly disclose its membership, although as recently as 2007 press releases have announced the names of those who have been invited to join. The 2007 release also stated that it has just under 6,000 voting members. While the membership had been growing, stricter policies have kept its size steady since then. "Academy Invites 115 to Become Members". Academy of Motion Picture Arts and Sciences. Archived from the original on August 27, 2007. . Retrieved September 4, 2007.In May 2011, the Academy sent a letter advising its 6,000 or so voting members that an online system for Oscar voting will be implemented in 2013.Cieply, Michael (May 23, 2011). "Electronic Voting Comes to The Oscars (Finally)". The New York Times. .Rules According to Rules 2 and 3 of the official Academy Awards Rules, a film must open in the previous calendar year, from midnight at the start of January 1 to midnight at the end of December 31, in Los Angeles County, California, to qualify (except for the Best Foreign Language Film). "Rule Two: Eligibility". Academy of Motion Picture Arts and Sciences. . Retrieved April 13, 2007. For example, the 2009 Academy Award for Best PictureBest Picture winner, The Hurt Locker, was actually first released in 2008, but did not qualify for the 81st Academy Awards2008 awards as it did not play its Oscar-qualifying run in Los Angeles until mid-2009, thus qualifying for the 82nd Academy Awards2009 awards.Rule 2 states that a film must be feature-length, defined as a minimum of 40 minutes, except for short subject awards, and it must exist either on a 35 mm film35 mm or 70 mm film print or in 24 frame/s or 48 frame/s progressive scan digital cinema format with native resolution not less than 720p1280x720. Producers must submit an Official Screen Credits online form before the deadline; in case it is not submitted by the defined deadline, the film will be ineligible for Academy Awards in any year. The form includes the production credits for all related categories. Then, each form is checked and put in a Reminder List of Eligible Releases. In late December ballots and copies of the Reminder List of Eligible Releases are mailed to around 6000 active members. For most categories, members from each of the branches vote to determine the nominees only in their respective categories (i.e. only directors vote for directors, writers for writers, actors for actors, etc.). There are some exceptions in the case of certain categories, like Foreign Film, Documentary and Animated Feature Film, in which movies are selected by special screening committees made up of members from all branches. In the special case of Best Picture, all voting members are eligible to select the nominees for that category. Foreign films must include English subtitles, and each country can submit only one film per year. "The Academy and its Oscar Awards – Reminder List of Eligible Releases". .The members of the various branches nominate those in their respective fields, while all members may submit nominees for Best Picture. The winners are then determined by a second round of voting in which all members are then allowed to vote in most categories, including Best Picture. "Rule Five: Balloting and Nominations". Academy of Motion Picture Arts and Sciences. . Retrieved April 13, 2007.Ceremony Telecast 31st Academy Awards Presentations, Pantages Theatre (Hollywood)Pantages Theater, Hollywood, 195981st Academy Awards Presentations, Kodak Theater, Hollywood, 2009 The major awards are presented at a live televised ceremony, most commonly in February or March following the relevant calendar year, and six weeks after the announcement of the nominees. It is the culmination of the film awards season, which usually begins during November or December of the previous year. This is an elaborate extravaganza, with the invited guests walking up the red carpet in the creations of the most prominent fashion designers of the day. Black tie dress is the most common outfit for men, although fashion may dictate not wearing a bow-tie, and musical performers sometimes do not adhere to this. (The artists who recorded the nominees for Best Original Song quite often perform those songs live at the awards ceremony, and the fact that they are performing is often used to promote the television broadcast). The Academy Awards is televised live across the United States (excluding Hawaii; they aired live for the first time in Alaska in 2011), Canada, the United Kingdom, and gathers millions of viewers elsewhere throughout the world. "International Broadcasters from Oscars.com". Oscars.com. . The 2007 ceremony was watched by more than 40 million Americans. Nielsen – Press Release: The Nielsen Company's 2008 Guide to the Academy Awards Other awards ceremonies (such as the Emmys, Golden Globes, and Grammy AwardGrammys) are broadcast live in the East Coast but are on Broadcast delaytape delay in the West Coast and might not air on the same day outside North America (if the awards are even televised). The Academy has for several years claimed that the award show has up to a billion viewers internationally, but this has so far not been confirmed by any independent sources. The Awards show was first televised on NBC in 1953. NBC continued to broadcast the event until 1960 when the American Broadcasting CompanyABC Network took over, televising the festivities through 1970, after which NBC resumed the broadcasts. ABC once again took over broadcast duties in 1976; it is under contract to do so through the year 2020. "ABC and Academy Extend Oscar Telecast Agreement" (Press release). Academy of Motion Picture Arts and Sciences. February 24, 2011. . Retrieved February 24, 2011.After more than 60 years of being held in late March or early April, the ceremonies were moved up to late February or early March starting in 2004 to help disrupt and shorten the intense For Your Consideration (advertising)lobbying and ad campaigns associated with Oscar season in the film industry. Another reason was because of the growing TV ratings success of the NCAA Men's Division I Basketball Championship, which would cut into the Academy Awards audience. The earlier date is also to the advantage of ABC, as it now usually occurs during the highly profitable and important February sweeps period. (Some years, the ceremony is moved into early March in deference to the Winter Olympics.) Advertising is somewhat restricted, however, as traditionally no movie studios or competitors of official Academy Award sponsors may advertise during the telecast. The Awards show holds the distinction of having won the most Emmys in history, with 47 wins and 195 nominations.Tom O'Neil (July 12, 2010). "Emmys love for Oscars continues with 12 nominations". Los Angeles Times. Archived from the original on August 19, 2010. . Retrieved August 13, 2010.After many years of being held on Mondays at 9:00 pm Eastern/6:00 p.m Pacific, in 1999 the ceremonies were moved to Sundays at 8:30 pm Eastern/5:30 pm Pacific.Bill Carter (April 8, 1998). "TV Notes; Moving Oscar Night". The New York Times. . Retrieved March 8, 2010. The reasons given for the move were that more viewers would tune in on Sundays, that Los Angeles rush-hour traffic jams could be avoided, and that an earlier start time would allow viewers on the East Coast to go to bed earlier. Academy Awards will move to Sunday night Reading Eagle – July 1, 1998; From Google News Archive For many years the film industry had opposed a Sunday broadcast because it would cut into the weekend box office. Never Say Never: Academy Awards move to Sunday The Item – March 19, 1999; From Google News ArchiveOn March 30, 1981, the awards ceremony was postponed for one day after Reagan assassination attemptthe shooting of President Ronald Reagan and others in Washington, D.C.In 1993, an In Memoriam segment was introduced,Child, Ben (March 10, 2010). "Farrah Fawcett:Oscars director apologises for 'In Memoriam' omission". The Guardian (London). Archived from the original on April 14, 2010. . Retrieved March 8, 2010. honoring those who had made a significant contribution to cinema who had died in the preceding 12 months, a selection compiled by a small committee of Academy members.Cohen, Sandy (March 3, 2010). "Oscar's 'In Memoriam' segment is touching to watch, painful to make". USA Today. Associated Press. . Retrieved March 8, 2010. This segment has drawn criticism over the years for the omission of some names.In 2010, the organizers of the Academy Awards announced that winners' acceptance speeches must not run past 45 seconds. This, according to organizer Bill Mechanic, was to ensure the elimination of what he termed "the single most hated thing on the show" – overly long and embarrassing displays of emotion.Jones, Sam (February 16, 2010). "Cut ... all change at Oscars as winners are given just 45 seconds to say thanks". The Guardian (London). .The Academy has also had recent discussions about moving the ceremony even further back into January, citing TV viewers' fatigue with the film industry's long awards season. But such an accelerated schedule would dramatically decrease the voting period for its members, to the point where some voters would only have time to view the contending films streamed on their computers (as opposed to traditionally receiving the films and ballots in the mail). Also, a January ceremony may have to compete with National Football League playoff games.John Horn (October 5, 2010). "Academy looks to move 2012 Oscar ceremony up several weeks". Los Angeles Times. . Retrieved February 28, 2011.Awards ceremonies The following is a listing of all Academy Awards ceremonies.Scott Bowles (February 26, 2008). "Low Oscar Ratings Cue Soul-Searching". USA Today. . Retrieved March 19, 2008.Nikki Finke (February 26, 2007). "UPDATE: 39.9 Million Watch 79th Oscars". Nikki Finke's Deadline Hollywood Daily (LA Weekly). . Retrieved February 21, 2011.Bill Gorman (March 8, 2010). "Academy Awards Averages 41.3 Million Viewers; Most Since 2005". TVbytheNumbers. Archived from the original on March 10, 2010. . Retrieved March 12, 2010.Ceremony Date Best Picture winner Length of ceremony Number of viewers Rating Host(s) Venue 1st Academy Awards May 16, 1929 Wings (film)Wings2 hours, 19 minutes — — Douglas Fairbanks, William C. deMilleHollywood Roosevelt Hotel2nd Academy Awards April 3, 1930 The Broadway Melody1 hour, 50 minutes — — William C. deMilleAmbassador Hotel (Los Angeles)Ambassador Hotel3rd Academy Awards November 5, 1930 All Quiet on the Western Front (1930 film)All Quiet on the Western Front2 hours, 13 minutes — — Conrad Nagel4th Academy Awards November 10, 1931Cimarron (1931 film)Cimarron2 hours, 3 minutes — — Lawrence GrantMillennium Biltmore HotelBiltmore Hotel5th Academy Awards November 18, 1932 Grand Hotel (film)Grand Hotel1 hour, 52 minutes — — Lionel Barrymore, Conrad NagelAmbassador Hotel (Los Angeles)Ambassador Hotel6th Academy Awards March 16, 1934 Cavalcade (1933 film)Cavalcade1 hour, 50 minutes — — Will Rogers7th Academy Awards February 27, 1935 It Happened One Night1 hour, 45 minutes — — Irvin S. CobbMillennium Biltmore HotelBiltmore Hotel8th Academy Awards March 5, 1936 Mutiny on the Bounty (1935 film)Mutiny on the Bounty2 hours, 12 minutes — — Frank Capra9th Academy Awards March 4, 1937 The Great Ziegfeld2 hours, 56 minutes — — George Jessel (actor)George Jessel10th Academy Awards March 10, 1938 The Life of Emile Zola1 hour, 56 minutes — — Bob Burns (comedian)Bob Burns11th Academy Awards February 23, 1939 You Can't Take It With You (film)You Can't Take It With You2 hours, 6 minutes — — None 12th Academy Awards February 29, 1940 Gone with the Wind (film)Gone with the Wind3 hours, 52 minutes — — Bob HopeAmbassador Hotel (Los Angeles)Ambassador Hotel (Cocoanut Grove) 13th Academy Awards February 27, 1941 Rebecca (1940 film)Rebecca2 hours, 10 minutes — — Bob HopeMillennium Biltmore HotelBiltmore Hotel (Biltmore Bowl) 14th Academy Awards February 26, 1942 How Green Was My Valley (film)How Green Was My Valley1 hour, 48 minutes — — None 15th Academy Awards March 4, 1943 Mrs. Miniver (film)Mrs. Miniver2 hours, 14 minutes — — Bob HopeAmbassador Hotel (Los Angeles)Ambassador Hotel (Cocoanut Grove) 16th Academy Awards March

2, 1944 Casablanca (film)Casablanca1 hour, 42 minutes — — Jack BennyGrauman's Chinese Theater17th Academy Awards March 15, 1945 Going My Way2 hours, 10 minutes — — Bob Hope, John Cromwell (director)John Cromwell18th Academy Awards March 7, 1946 The Lost Weekend (film)The Lost Weekend1 hour, 41 minutes — — Bob Hope, James Stewart19th Academy Awards March 13, 1947 The Best Years of Our Lives2 hours, 52 minutes — — Jack BennyShrine Auditorium20th Academy Awards March 20, 1948 Gentleman's Agreement1 hour, 58 minutes — — Agnes Moorehead, Dick Powell21st Academy Awards March 24, 1949 Hamlet (1948 film)Hamlet1 hour, 35 minutes — — Robert Montgomery (actor)Robert MontgomeryThe Academy Theater22nd Academy Awards March 23, 1950 All the King's Men (1949 film)All the King's Men1 hour, 50 minutes — — Paul Douglas (actor)Paul DouglasPantages Theatre (Hollywood)Pantages Theatre23rd Academy Awards March 29, 1951 All About Eve2 hours, 18 minutes — — Fred Astaire24th Academy Awards March 20, 1952 An American in Paris (film)An American in Paris1 hour, 53 minutes — — Danny Kaye25th Academy Awards March 19, 1953 The Greatest Show on Earth1 hour, 32 minutes 40 million — Bob Hope, Conrad NagelPantages Theatre (Hollywood)Pantages Theatre / NBC International Theatre 26th Academy Awards March 25, 1954 From Here to Eternity1 hour, 58 minutes 43 million — Donald O'Connor, Fredric MarchPantages Theatre (Hollywood)Pantages Theatre / New Century TheatreNBC Century Theatre27th Academy Awards March 30, 1955 On the Waterfront1 hour, 48 minutes — — Bob Hope, Thelma Ritter28th Academy Awards March 21, 1956 Marty (film)Marty1 hour, 30 minutes — — Jerry Lewis, Claudette Colbert, Joseph L. Mankiewicz29th Academy Awards March 27, 1957 Around the World in Eighty Days (1956 film)Around the World in 80 Days3 hours, 8 minutes — — Jerry Lewis, Celeste Holm30th Academy Awards March 26, 1958 The Bridge on the River Kwai2 hours, 41 minutes — — Bob Hope, David Niven, James Stewart, Jack Lemmon, Rosalind RussellPantages Theatre (Hollywood)Pantages Theatre31st Academy Awards April 6, 1959 Gigi (1958 film)Gigi1 hour, 55 minutes — — Bob Hope, David Niven, Tony Randall, Mort Sahl, Laurence OliverSir Laurence Olivier, Jerry Lewis32nd Academy Awards April 4, 1960 Ben-Hur (1959 film)Ben-Hur1 hour, 40 minutes — — Bob Hope33rd Academy Awards April 17, 1961 The Apartment2 hours, 5 minutes — — Bob HopeSanta Monica Civic Auditorium34th Academy Awards April 9, 1962 West Side Story (film)West Side Story2 hours, 10 minutes — — Bob Hope35th Academy Awards April 8, 1963 Lawrence of Arabia (film)Lawrence of Arabia2 hours, 30 minutes — — Frank Sinatra36th Academy Awards April 13, 1964 Tom Jones (film)Tom Jones2 hours, 8 minutes — — Jack Lemmon37th Academy Awards April 5, 1965 My Fair Lady (film)My Fair Lady2 hours, 50 minutes — — Bob Hope38th Academy Awards April 18, 1966 The Sound of Music (film)The Sound of Music2 hours, 54 minutes — — Bob Hope39th Academy Awards April 10, 1967 A Man for All Seasons (1966 film)A Man for All Seasons2 hours, 31 minutes — — Bob Hope40th Academy Awards April 10, 1968 In the Heat of the Night (film)In the Heat of the Night1 hour, 50 minutes — — Bob Hope41st Academy Awards April 14, 1969 Oliver! (film)Oliver!2 hours, 33 minutes — — None Dorothy Chandler Pavilion42nd Academy Awards April 7, 1970 Midnight Cowboy2 hours, 25 minutes — 43.40 None 43rd Academy Awards April 15, 1971 Patton (film)Patton2 hours, 52 minutes — — None 44th Academy Awards April 10, 1972 The French Connection (film)The French Connection1 hour, 44 minutes — — Helen Hayes, Alan King (comedian)Alan King, Sammy Davis, Jr., Jack Lemmon45th Academy Awards March 27, 1973 The Godfather2 hours, 38 minutes — — Carol Burnett, Michael Caine, Charlton Heston, Rock Hudson46th Academy Awards April 2, 1974 The Sting3 hours, 23 minutes — — John Huston, Burt Reynolds, David Niven, Diana Ross47th Academy Awards April 8, 1975 The Godfather Part II3 hours, 20 minutes — — Sammy Davis, Jr., Bob Hope, Shirley MacLaine, Frank Sinatra48th Academy Awards March 29, 1976 One Flew Over the Cuckoo's Nest (film)One Flew Over the Cuckoo's Nest3 hours, 12 minutes — — Goldie Hawn, Gene Kelly, Walter Matthau, George Segal, Robert Shaw (British actor)Robert Shaw49th Academy Awards March 28, 1977 Rocky3 hours, 38 minutes — — Warren Beatty, Ellen Burstyn, Jane Fonda, Richard Pryor50th Academy Awards April 3, 1978 Annie Hall3 hours, 30 minutes39.73 million 31.10 Bob Hope51st Academy Awards April 9, 1979 The Deer Hunter3 hours, 25 minutes — — Johnny Carson52nd Academy Awards April 14, 1980 Kramer vs. Kramer3 hours, 12 minutes — — Johnny Carson53rd Academy Awards March 31, 1981 Ordinary People3 hours, 13 minutes — — Johnny Carson54th Academy Awards March 29, 1982 Chariots of Fire3 hours, 24 minutes — — Johnny Carson55th Academy Awards

April 11, 1983 Gandhi (film)Gandhi3 hours, 15 minutes — — Liza Minnelli, Dudley Moore, Richard Pryor, Walter Matthau56th Academy Awards April 9, 1984 Terms of Endearment3 hours, 42 minutes — 38.00 Johnny Carson57th Academy Awards March 25, 1985 Amadeus (film)Amadeus3 hours, 10 minutes — — Jack Lemmon58th Academy Awards March 24, 1986 Out of Africa (film)Out of Africa3 hours, 2 minutes 38.65 million 25.71 Alan Alda, Jane Fonda, Robin Williams59th Academy Awards March 30, 1987 Platoon (film)Platoon3 hours, 19 minutes 39.72 million 25.94 Chevy Chase, Goldie Hawn, Paul Hogan60th Academy Awards April 11, 1988 The Last Emperor3 hours, 33 minutes 42.04 million 27.80 Chevy ChaseShrine Auditorium61st Academy Awards March 29, 1989 Rain Man3 hours, 19 minutes 42.77 million 28.41 None 62nd Academy Awards March 26, 1990 Driving Miss Daisy3 hours, 37 minutes 40.22 million 26.42 Billy CrystalDorothy Chandler Pavilion63rd Academy Awards March 25, 1991 Dances with Wolves3 hours, 35 minutes 42.79 million 28.06 Billy CrystalShrine Auditorium64th Academy Awards March 30, 1992 The Silence of the Lambs (film)The Silence of the Lambs3 hours, 33 minutes 44.44 million 29.84 Billy CrystalDorothy Chandler Pavilion65th Academy Awards March 29, 1993 Unforgiven3 hours, 30 minutes 45.84 million 32.85 Billy Crystal66th Academy Awards March 21, 1994 Schindler's List3 hours, 18 minutes 46.26 million 31.86 Whoopi Goldberg67th Academy Awards March 27, 1995 Forrest Gump3 hours, 35 minutes 48.87 million 33.47 David LettermanShrine Auditorium68th Academy Awards March 25, 1996 Braveheart3 hours, 38 minutes 44.81 million 30.48 Whoopi GoldbergDorothy Chandler Pavilion69th Academy Awards March 24, 1997 The English Patient (film)The English Patient3 hours, 34 minutes 40.83 million 25.83 Billy CrystalShrine Auditorium70th Academy Awards March 23, 1998 Titanic (1997 film)Titanic3 hours, 47 minutes 57.25 million 35.32 Billy Crystal71st Academy Awards March 21, 1999 Shakespeare in Love4 hours, 2 minutes 45.63 million 28.51 Whoopi GoldbergDorothy Chandler Pavilion72nd Academy Awards March 26, 2000 American Beauty (film)American Beauty4 hours, 4 minutes 46.53 million 29.64 Billy CrystalShrine Auditorium73rd Academy Awards March 25, 2001 Gladiator (2000 film)Gladiator3 hours, 23 minutes 42.93 million 25.86 Steve Martin74th Academy Awards March 24, 2002 A Beautiful Mind (film)A Beautiful Mind4 hours, 23 minutes 40.54 million 25.13 Whoopi GoldbergDolby Theatre75th Academy Awards March 23, 2003 Chicago (2002 film)Chicago3 hours, 30 minutes 33.04 million 20.58 Steve Martin76th Academy Awards February 29, 2004 The Lord of the Rings: The Return of the KingThe Lord of the Rings:The Return of the King3 hours, 44 minutes 43.56 million 26.68 Billy Crystal77th Academy Awards February 27, 2005 Million Dollar Baby3 hours, 14 minutes 42.16 million 25.29 Chris Rock78th Academy Awards March 5, 2006 Crash (2004 film)Crash3 hours, 33 minutes 38.64 million 22.91 Jon Stewart79th Academy Awards February 25, 2007 The Departed3 hours, 51 minutes 39.92 million 23.65 Ellen DeGeneres80th Academy Awards February 24, 2008 No Country for Old Men (film)No Country for Old Men3 hours, 21 minutes 31.76 million 18.66 Jon Stewart81st Academy Awards February 22, 2009 Slumdog Millionaire3 hours, 30 minutes 36.94 million 21.68 Hugh Jackman82nd Academy Awards March 7, 2010 The Hurt Locker3 hours, 37 minutes 41.62 million 24.75 Steve Martin, Alec Baldwin83rd Academy Awards February 27, 2011 The King's Speech3 hours, 15 minutes 37.63 million 21.97 James Franco, Anne Hathaway84th Academy Awards February 26, 2012 The Artist (film)The Artist3 hours, 14 minutes 39.30 million 25.50 Billy Crystal85th Academy Awards February 24, 2013 Academy Sets Date for Next Year's Oscars I The Wrap Awards — — — — Seth MacFarlane "SETH MACFARLANE TO HOST 85TH OSCARS®, AIRING LIVE ON ABC, SUNDAY, FEBRUARY 24, 2013". ABC Media Net. American Broadcasting Company. October 1, 2012. . Retrieved October 1, 2012.Ceremony Date Best Picture winner Length of ceremony Number of viewers Rating Host(s) Venue Historically, the "Oscarcast" has pulled in a bigger haul when box-office hits are favored to win the Best Picture trophy. More than 57.25 million viewers tuned to the telecast for the 70th Academy Awards in 1998, the year of Titanic (1997 film)Titanic, which generated close to US$600 million at the North American box office pre-Oscars.James, Meg (February 23, 2008). "Academy's red carpet big stage for advertisers". The Seattle Times. . The 76th Academy Awards ceremony in which The Lord of the Rings: The Return of the King (pre-telecast box office earnings of US$368 million) received 11 Awards including Best Picture drew 43.56 million viewers.Bowles, Scott (January 26, 2005). "Oscars lack blockbuster to lure TV viewers". USA Today. . Retrieved November 8, 2006. The most watched ceremony based on Nielsen ratings to date, however, was the 42nd Academy Awards (Best

Picture Midnight Cowboy) which drew a 43.4% household rating on April 7, 1970. Charts and Data: Top 100 TV Shows of All Time by VarietyBy contrast, ceremonies honoring films that have not performed well at the box office tend to show weaker ratings. The 78th Academy Awards which awarded low-budgeted, independent film Crash (2004 film)Crash (with a pre-Oscar gross of US$53.4 million) generated an audience of 38.64 million with a household rating of 22.91%.Levin, Gary (March 7, 2006). "Low Ratings Crash Party". USA Today. . Retrieved April 14, 2010. In 2008, the 80th Academy Awards telecast was watched by 31.76 million viewers on average with an 18.66% household rating, the lowest rated and least watched ceremony to date, in spite of celebrating 80 years of the Academy Awards. "Oscar ratings worst ever". The Washington Post. . The Best Picture winner of that particular ceremony was another independently financed film (No Country for Old Men (film)No Country for Old Men).Venues Pantages Theatre (Hollywood)Pantages Theatre, 2008 In 1929, the first Academy Awards were presented at a banquet dinner at the Roosevelt Hotel (Hollywood)Hollywood Roosevelt Hotel. From 1930–1943, the ceremony alternated between two venues: the Ambassador Hotel (Los Angeles)Ambassador Hotel on Wilshire Boulevard and the Millennium Biltmore HotelBiltmore Hotel in downtown Los Angeles. Grauman's Chinese Theater in Hollywood then hosted the awards from 1944 to 1946, followed by the Shrine Auditorium in Los Angeles from 1947 to 1948. The 21st Academy Awards in 1949 were held at the Academy Award Theater at what was the Academy's headquarters on Melrose Avenue in Hollywood. "Oscars Award Venues". Academy of Motion Picture Arts and Sciences. Archived from the original on December 12, 2006. . Retrieved April 13, 2007.From 1950 to 1960, the awards were presented at Hollywood's Pantages Theatre (Hollywood)Pantages Theatre. With the advent of television, the 1953–1957 awards took place simultaneously in Hollywood and New York first at the NBC International Theatre (1953) and then at the NBC Century Theatre (1954–1957), after which the ceremony took place solely in Los Angeles. The Oscars moved to the Santa Monica Civic Auditorium in Santa Monica, California in 1961. By 1969, the Academy decided to move the ceremonies back to Los Angeles, this time to the Dorothy Chandler Pavilion at the Los Angeles Music CenterLos Angeles County Music Center. In 2002, the Hollywood and Highland Center (theatre)Kodak Theatre became the permanent home of the award ceremonies. However, due to Eastman Kodak's bankruptcy issues, this theatre was renamed the Hollywood and Highland Center (theatre)Hollywood and Highland Center in the days preceding the February 26, 2012, awards ceremony. As of May 2012, the theatre was once again renamed – to the Dolby Theatre – after Dolby Laboratories acquired the naming rights.Derrick J. Lang (May 12, 2012). "Oscars venue reopens as Dolby Theatre". Associated Press. . Retrieved May 12, 2012.Merit categories Current categories Academy Award for Best ActorBest Actor in a Leading Role: since 1928 Academy Award for Best Supporting ActorBest Actor in a Supporting Role: since 1936 Academy Award for Best ActressBest Actress in a Leading Role: since 1928 Academy Award for Best Supporting ActressBest Actress in a Supporting Role: since 1936 Academy Award for Best Animated FeatureBest Animated Feature: since 2001 Academy Award for Best Animated Short FilmBest Animated Short Film: since 1931 Academy Award for Best CinematographyBest Cinematography: since 1928 Academy Award for Best Costume DesignBest Costume Design: since 1948 Academy Award for Best DirectorBest Director: since 1928 Academy Award for Best DocumentaryBest Documentary Feature: since 1943 Academy Award for Best Documentary ShortBest Documentary Short: since 1941 Academy Award for Best Film EditingBest Film Editing: since 1935 Academy Award for Best Foreign Language FilmBest Foreign Language Film: since 1947 Academy Award for Best Live Action Short FilmBest Live Action Short Film: since 1931 Academy Award for Best Makeup and HairstylingBest Makeup and Hairstyling: since 1981 Academy Award for Best Original ScoreBest Original Score: since 1934 Academy Award for Best Original SongBest Original Song: since 1934 Academy Award for Best PictureBest Picture: since 1928 Academy Award for Best Production DesignBest Production Design: since 1928 Academy Award for Best Sound EditingBest Sound Editing: since 1963 Academy Award for Best SoundBest Sound Mixing: since 1930 Academy Award for Best Visual EffectsBest Visual Effects: since 1939 Academy Award for Best Adapted ScreenplayBest Adapted Screenplay: since 1928 Academy Award for Best Original ScreenplayBest Original Screenplay: since 1940 In the first year of the awards, the Best Director award was split into two separate categories (Drama and Comedy). At times, the Best Original Score award has also been split into separate categories (Drama and Comedy/Musical).

From the 1930s through the 1960s, the Art Direction (now Production Design), Cinematography, and Costume Design awards were likewise split into two separate categories (black-and-white films and color films). Prior to 2012, the Production Design award was called Art Direction, while the Makeup and Hairstyling award was called Makeup. Another award, entitled the Academy Award for Best Original Musical, is still in the Academy rulebooks and has yet to be discontinued. However, due to continuous insufficient eligibility each year, it has not been awarded since 1984 (when Purple Rain (film)Purple Rain won). Music Awards | Rules for the 84th Academy Awards | Academy of Motion Picture Arts & SciencesDiscontinued categories Academy Award for Best Assistant DirectorBest Assistant Director: 1933 to 1937 Academy Award for Best DirectorBest Director, Comedy Picture: 1928 only Academy Award for Best Dance DirectionBest Dance Direction: 1935 to 1937 Academy Award for Best Engineering EffectsBest Engineering Effects: 1928 only Academy Award for Best Original Musical or Comedy ScoreBest Original Musical or Comedy Score: 1995 to 1999 Academy Award for Best StoryBest Original Story: 1928 to 1956 Academy Award for Best Score – Adaptation or TreatmentBest Score – Adaptation or Treatment: 1962 to 1969; 1973 Academy Award for Best Short Film - ColorBest Short Film – Color: 1936 and 1937 Academy Award for Best Short Film - Live Action - 2 ReelsBest Short Film – Live Action – 2 Reels: 1936 to 1956 Academy Award for Best Short Film - NoveltyBest Short Film – Novelty: 1932 to 1935 Academy Award for Best Title WritingBest Title Writing: 1928 only Academy Award for Best Unique and Artistic ProductionBest Unique and Artistic Quality of Production: 1928 only Proposed categories The Board of Governors meets each year and considers new award categories. To date, the following proposed categories have been rejected: Best Casting: rejected in 1999 Best Stunt Coordination: rejected in 1999; rejected in 2005;Michael Hiltzik (August 4, 2005). "One stunt they've been unable to pull off". Los Angeles Times. . Retrieved April 13, 2007. rejected in 2011Handel, Jonathan (June 15, 2011). "Academy Votes Against Creating Oscar Category for Stunt Coordinators". The Hollywood Reporter. . Best Title Design: rejected in 1999 Special categories The Special Academy Awards are voted on by special committees, rather than by the Academy membership as a whole. They are not always presented on a consistent annual basis. Current special categories Academy Honorary Award: since 1929 Academy Scientific and Technical Award: since 1931 Gordon E. Sawyer Award: since 1981 Jean Hersholt Humanitarian Award: since 1956 Irving G. Thalberg Memorial Award: since 1938 Discontinued special categoriesAcademy Juvenile Award: 1934 to 1960 Special Achievement Academy AwardAcademy Special Achievement Award: 1972 to 1995 Parodies In 1987, at 25 years, has a parodied version Colombian called Hétores Awards (in Spanish Premios Hétores) given by the University Student Journalism, Advertising, among others. Rewarding the best parodies of Hollywood movies, Colombian TV series, music videos, among others. "Regresan los Hétores". Eltiempo.com. . Retrieved September 24, 2004. "Se acerca la entrega de los Hétores". Eltiempo.com. . Retrieved October 24, 1996. This event was born out of The Academy Awards, held annually in the city of Medellin and is transmitted through the TV channels of local or regional department of Antioquia. It has red carpet where the stars spend Colombian and parodies actors (producers), a Show Abroad (Where launches new Colombian artist), the awards ceremony (which has a main show of the biggest names in music , an area of Humor, and a time to showing the concept of the year). "Regresan los Hétores". Eltiempo.com. . Retrieved September 24, 2004.Criticism Due to the positive exposure and prestige of the Academy Awards, studios spend millions of dollars and hire publicists specifically to promote their films during what is typically called the "Oscar season". This has generated accusations of the Academy Awards being influenced more by marketing than quality. William Friedkin, an Academy Award-winning film director and former producer of the ceremony, expressed this sentiment at a conference in New York in 2009, describing it as "the greatest promotion scheme that any industry ever devised for itself".Friedkin, William (Director) (February 24, 2009). Director William Friedkin at the Hudson Union Society. Archived from the original on March 10, 2009. . Retrieved March 11, 2009.In addition, some winners critical of the Academy Awards have boycotted the ceremonies and refused to accept their Oscars. The first to do so was Dudley Nichols (Best Writing in 1935 for The Informer (1935 film)The Informer). Nichols boycotted the 8th Academy Awards ceremony because of conflicts between the Academy and the Writers' Guild. "The Oscars Did You Know?". Archived from the original on June 23, 2009. . Retrieved June 18, 2009. George C. Scott became the second person to refuse his award (Best Actor in 1970 for Patton (film)Patton) at the 43rd

Academy Awards ceremony. Scott described it as a 'meat parade', saying 'I don't want any part of it." "George C Scott: The man who refused an Oscar". BBC News. September 23, 1999. . "Show Business: Meat Parade". Time. March 8, 1971. . "Fast Facts – Did You Know?". Biography.com. May 16, 1929. Archived from the original on February 10, 2010. . Retrieved February 6, 2010. The third winner, Marlon Brando, refused his award (Best Actor in 1972 for The Godfather), citing the film industry's discrimination and mistreatment of Native Americans. At the 45th Academy Awards ceremony, Brando sent Sacheen Littlefeather to read a 15-page speech detailing his criticisms.Tim Dirks, editor of American Movie ClassicsAMC's filmsite.org, has written of the Academy Awards, Unfortunately, the critical worth, artistic vision, cultural influence, and innovative qualities of many films are not given the same voting weight. Especially since the 1980s, moneymaking "formula-made" blockbusters with glossy production values have often been crowd-pleasing titans (and Best Picture winners), but they haven't necessarily been great films with depth or critical acclaim by any measure. "Academy Awards – The Oscars". . Retrieved October 4, 2009.Acting prizes in certain years have been criticized for not recognizing superior performances so much as being awarded for sentimental reasons, "Taylor, Elizabeth". . Retrieved October 4, 2009. personal popularity, "What's the worst Best Actor choice of all time?". . Retrieved October 4, 2009. atonement for past mistakes, "Being an Oscar voter *doesn't* mean never having to say you're sorry". The Los Angeles Times. . Retrieved October 4, 2009. or presented as a "career honor" to recognize a distinguished nominee's entire body of work. All about Oscar: the history and politics of the Academy Awards – The Career Oscars. 2003. ISBN 978-0-8264-1452-6. . Retrieved October 4, 2009.Associated events The following events are closely associated with the annual Academy Awards ceremony: Nominees luncheon Governors Awards The 25th Independent Spirit Awards (in 2010), usually held in Santa Monica the Saturday before the Oscars, marked the first time it was moved to a Friday and a change of venue to L.A. Live. Golden Raspberry Awards The 8th annual "Night Before", traditionally held at the Beverly Hills Hotel (eight years running in 2010) and generally known as THE party of the season, benefits the Motion Picture and Television Fund, which operates a retirement home for SAG actors in the San Fernando Valley.Elton John AIDS Foundation Academy Award Party airs the awards live at the nearby Pacific Design Center. The Governors' Ball is the Academy's official after-party, including dinner (until 2011), and is held adjacent to the awards-presentation venue. In 2012, the three course meal was replaced by appetizers. The Vanity Fair after-party, historically held at the former Morton's restaurant, since 2009 has been held at the Sunset Tower. See also Academy Awards pre-showElton John AIDS Foundation Academy Award PartyGolden Raspberry AwardOscar seasonCannes Film FestivalCamerimageBerlin International Film FestivalVenice Film FestivalSan Sebastián International Film FestivalLists:List of Academy Award recordsList of Academy Award winning familiesList of Academy Award-winning filmsList of Academy Awards ceremoniesList of Academy Award trophies on public displayList of actors who have appeared in multiple Best Picture Academy Award winnersList of Big Five Academy Award winners and nomineesList of fictitious Academy Award winners and nomineesList of foreign language films nominated for Academy AwardsList of oldest and youngest Academy Award winners and nomineesList of people who have won multiple Academy Awards in a single yearList of posthumous Academy Award winners and nomineesList of presenters of Best Picture Academy AwardList of Best Picture MilestonesList of superlative Academy Award winners and nomineesNationality:List of Australian Academy Award winners and nomineesList of British and Commonwealth Academy Award winners and nomineesList of English Academy Award nominees and winnersList of French Academy Award winners and nomineesList of Italian Academy Award winners and nomineesList of Latin American Academy Award winners and nomineesList of Polish Academy Award winners and nomineesList of Scandinavian Academy Award winners and nomineesList of Spanish Academy Award winners and nomineesRace:List of Asian Academy Award winners and nomineesList of Black Academy Award winners and nomineesLists of Hispanic Academy Award winners and nominees by countryGeneral:British Academy Film AwardsBAFTA (British equivalent) César Award (French equivalent) Premios GoyaGoya Awards (Spanish equivalent) List of years in filmNotes References Brokaw, Lauren (2010). "Wanna see an Academy Awards invite? We got it along with all the major annual events surrounding the Oscars". Los Angeles: The Daily Truffle. Cotte, Oliver (2007). Secrets of Oscar-winning animation: Behind the scenes of 13 classic short animations. Focal Press.

ISBN 978-0-240-52070-4. Gail, K., and Piazza, J. (2002). The Academy Awards: The Complete History of Oscar. Black Dog & Leventhal Publishers, Inc. ISBN 1-57912-240-X. Levy, Emanuel (2003). All About Oscar: The History and Politics of the Academy Awards. New York: Continuum. ISBN 0-8264-1452-4. Wright, Jon (2007). The Lunacy of Oscar: The Problems with Hollywood's Biggest Night. Thomas Publishing, Inc.External links Official website Oscar.com—official Academy Award ceremony site Academy Awards at the Open Directory Project "Oscar Greats" at Time (magazine)Time magazine Academy Awards USA at IMDB

Fiorello_La_Guardia

Fiorello La Guardia		
[[file:Fiorello LaGuardia.jpg		alt=]]
99th Mayor of New York City[1]		
In office January 1, 1934 – December 31, 1945		
Preceded by	John P. O'Brien	
Succeeded by	William O'Dwyer	
Member of the U.S. House of Representatives from New York's 20th district		
In office March 4, 1923 – March 3, 1933		
Preceded by	Isaac Siegel	
Succeeded by	James J. Lanzetta	
10th President of the New York City Board of Aldermen		
In office January 1, 1920 – December 31, 1921		
Preceded by	Robert L. Moran	
Succeeded by	Murray Hulbert	
Member of the U.S. House of Representatives from New York's 14th district		
In office March 4, 1917 – December 31, 1919		
Preceded by	Michael F. Farley	
Succeeded by	Nathan D. Perlman	
Personal details		
Born	Fiorello Enrico LaGuardiaDecember 11, 1882Greenwich Village, Manhattan, New York, United States	
Died	September 20, 1947 (aged 64)Bronx, New York, United States	
Political party	Republican	
Profession	Politician	
Religion	Episcopalian	
Signature	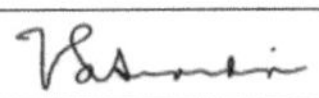	

Fiorello Henry LaGuardia (pron.: /fiəˈrɛloʊləˈɡwɑrdiə/) (born **Fiorello Enrico La Guardia**)[2] (December 11, 1882 – September 20, 1947) was Mayor of New York for three terms from 1934 to 1945 as a Republican. Previously he was elected to Congress in 1916 and 1918, and again from 1922 through 1930. Irascible, energetic and charismatic, he craved publicity and is acclaimed as one of the three or four greatest mayors in American history.[3] Only five feet tall, he was called "the Little Flower" (*Fiorello* is Italian for "little flower").

LaGuardia, a Republican who appealed across party lines, was very popular in New York during the 1930s. As a New Dealer, he supported President Franklin D. Roosevelt, a Democrat, and in turn Roosevelt heavily funded the city and cut off patronage from LaGuardia's foes. La Guardia revitalized New York City and restored public faith in

City Hall. He unified the transit system; directed the building of low-cost public housing, public playgrounds, and parks; constructed airports; reorganized the police force; defeated the powerful Tammany Hall political machine; and reestablished merit employment in place of patronage jobs.[4]

LaGuardia was a domineering leader who verged on authoritarianism but whose reform politics were carefully tailored to address the sentiments of his diverse constituency. He defeated a corrupt Democratic machine, presided during a depression and a world war, made the city the model for New Deal welfare and public works programs, and championed immigrants and ethnic minorities. He succeeded with the support of a sympathetic president. He secured his place in history as a tough-minded reform mayor who helped clean out corruption, bring in gifted experts, and fix upon the city a broad sense of responsibility for its own citizens. His administration engaged new groups that had been kept out of the political system, gave New York its modern infrastructure, and raised expectations of new levels of urban possibility.

The intemperate mayor was rough on his staffers and left no doubt who was in charge. He lost his intuitive touch during the war years, when the federal money stopped flowing in, and never realized that he had created far more infrastructure than the city could afford. "LaGuardia represented a dangerous style of personal rule hitched to a transcendent purpose," according to Thomas Kessner, LaGuardia's biographer. "People would be afraid of allowing anybody to take that kind of power today."[4] [5]

Early life and career

LaGuardia was born in Greenwich Village in New York City to two Italian immigrant parents. His father, Achille La Guardia, was a lapsed-Catholic from Cerignola, and his mother, Irene Coen, was a Jew from Trieste, then part of the Austro-Hungarian Empire; his maternal grandmother Fiorina Luzzatto Coen was a Luzzatto, a member of the prestigious Italian Jewish family of scholars, kabbalists and poets and had among her ancestors the famous rabbi Samuel David Luzzatto. It was in Trieste that Achille La Guardia met and married Irene .[6] Fiorello La Guardia was raised an Episcopalian and practised that religion all his life. His middle name “Enrico” was changed to “Henry” (the English form of Enrico) when he was a child.

He moved to Arizona with his family, where his father had a bandmaster position at Fort Whipple in the U.S. Army. LaGuardia attended public schools and high school in Prescott, Arizona.[7] After his father was discharged from his bandmaster position in 1898, Fiorello lived in Trieste.[8]

La Guardia joined the State Department and served in U.S. consulates in Budapest, Trieste (Italy), and Fiume (Austria-Hungary), now Rijeka (Croatia), (1901–1906). He returned to the U.S. to continue his education at New York University. In 1907–10, he worked for New York Society for the Prevention of Cruelty to Children as an interpreter for the U.S. Bureau of Immigration at the Ellis Island immigrant station.

He graduated from New York University School of Law in 1910, was admitted to the bar the same year, and began a law practice in New York City.[7]

Marriages and family

LaGuardia married twice. His first wife was Thea Almerigotti, whom he married on March 8, 1919. In November 1920 they had a daughter, Fioretta Thea, who died May 8, 1921 of spinal meningitis. His wife died of tuberculosis on November 29, 1921, at the age of 26.[9] He married Marie Fisher in 1929; they adopted two children.

Early political career

Fiorello LaGuardia between two Italian officers in front of a Ca.44, c. 1918

Elected to Congress

LaGuardia became Deputy Attorney General of New York in January, 1915.[10] In 1916, he was elected to the U.S. House of Representatives, where he had a reputation as a fiery and devoted reformer. As a congressman, LaGuardia represented an ethnically diverse slum district in East Harlem and, although barred from important committee posts because of his political independence, he was a tireless and vocal champion of Progressive causes.[11] LaGuardia took office on March 4, 1917 but soon was commissioned in the United States Army Air Service ; he rose to the rank of major in command of a unit of Ca.44 bombers on the Italian-Austrian front in World War I. LaGuardia resigned his seat in Congress on December 31, 1919.

President of the Board of Aldermen

In 1919, LaGuardia was chosen to run as the Republican candidate for the office of President of the New York City Board of Aldermen. His Democratic opponent was Robert L. Moran, an alderman from the Bronx who had succeeded to the Board presidency in 1918 when Alfred E. Smith, who had been elected Board President in 1917, became Governor.[12] Michael "Dynamite Mike" Kelly, commander of New York's Third "Shamrock" Battalion, also joined the race. Tammany Hall looked with alarm upon Kelly's entrance into the campaign and tried to persuade him to withdraw his candidacy and throw his support behind Moran. When he refused, Tammany went to the New York Supreme Court and successfully sued to keep Kelly's name off the ballot.[13] When Election Day arrived, over 3,500 of Kelly's supporters wrote LaGuardia's name on the ballot.[13] This number was sufficient to defeat Moran, who lost to LaGuardia by only 1,363 votes.[14]

LA GUARDIA WINS BY 1,530

Beats Moran for President of Board of Aldermen in a Close Contest.

KOENIG ORDERED VIGILANCE

Warned Republican Chairmen to Stay by the Ballot Boxes and Scrutinize Count.

CURRAN DEFEATS BOYLE

Five Republican Votes in Board of Estimate Assured—Clean-Cut Result in Supreme Court.

New York Times front page November 5, 1919

Return to Congress

LaGuardia, running as a Republican, won a seat in Congress from the Italian stronghold of East Harlem in 1922 and served in the House until March 3, 1933.[11] A leading liberal reformer, LaGuardia sponsored labor legislation and railied against immigration quotas. His major legislation was the Norris-LaGuardia Act, cosponsored with Nebraska senator George Norris in 1932. It circumvented Supreme Court limitations on the activities of labor unions, especially as those limitations were imposed between the enactment of the Clayton Antitrust Act in 1914 and the end of the 1920s. Based on the theory that the lower courts are creations not of the Constitution but of Congress, and that Congress therefore has wide power in defining and restricting their jurisdiction, the act forbids issuance of injunctions to sustain anti-union contracts of employment, to prevent ceasing or refusing to perform any work or remain in any relation of employment, or to restrain acts generally constituting component parts of strikes, boycotts, and picketing. It also said courts could no longer enforce yellow-dog contracts, which are labor contracts prohibiting a worker from joining a union.[15] [16]

LaGuardia during his time in Congress, c. 1929.

Foreign policy

Never an isolationist, he supported using American influence abroad on behalf of democracy or for national independence or against autocracy. Thus he supported the Irish independence movement and the anti-czarist Russian Revolution of 1917, but did not approve of Vladimir Lenin. Unlike most progressive colleagues, such as Norris, La Guardia consistently backed internationalism, speaking in favor of the League of Nations and the Inter-Parliamentary Union as well as peace and disarmament conferences. In domestic policies he tended toward socialism and wanted to nationalize and regulate; he was never close to the Socialist Party and never bothered to read Karl Marx.[17]

Champion of the progressive movement

As a congressman, LaGuardia was a tireless and vocal champion of progressive causes, from allowing more immigration and removing U.S. troops from Nicaragua to speaking up for the rights and livelihoods of striking miners, impoverished farmers, oppressed minorities, and struggling families. A goad to the era's plutocrats and their enablers in government, LaGuardia fought for progressive income taxes, greater government oversight of Wall Street, and national employment insurance for workers idled by the Great Depression.[18]

Defeats in 1929 and 1932

As a Republican La Guardia had to support Harding in 1920; he had to be silent in the 1928 campaign although he favored Al Smith, a Democrat. In 1929, he lost the election for mayor to incumbent Democrat Jimmy Walker by a landslide.[19] In 1932, he was defeated for re-election to the House by James J. Lanzetta, the Democratic candidate. 1932 was not a good year for Republican candidates like LaGuardia, and the 20th Congressional district was shifting from a Jewish and Italian-American population to a Puerto Rican population. However, it has also been argued that powerful Tammany Hall boss Jimmy Hines was able to successfully get enough votes forged to get LaGuardia unseated in this election as well.[20]

Mayor of New York

1933 election

Walker and his Irish-run Tammany Hall were forced out of office by scandal and LaGuardia was determined to replace him. First he had to win the nomination of both the Republican party and also the "Fusion" group of independents. He was not the first choice of either, for they distrusted Italians. On the other hand La Guardia had enormous determination, high visibility, the support of reformer Samuel Seabury and the ability to ruin prospects of any rival by a divisive primary contest. He secured the nominations and expected an easy win against hapless incumbent Mayor John P. O'Brien. At the last minute Joseph V. McKee entered the race as the nominee of the new "Recovery party." McKee was a formidable opponent because he was sponsored by Bronx Democratic boss Edward J. Flynn and apparently was favored by President Franklin Roosevelt. LaGuardia made corruption his main issue. The campaign saw mud slung three ways, with LaGuardia denounced as a far-left "Red," O'Brien as a pawn of the bosses, and McKee as an anti-Semite. LaGuardia's win was based on a complex coalition of regular Republicans (mostly middle class Germans in the boroughs outside Manhattan), a minority of reform-minded Democrats, some Socialists, a large proportion of middle-class Jews, and the great majority of Italians. The Italians had been loyal to Tammany; their switch proved decisive.[21]

Agenda

LaGuardia came to office in January 1934 with five main goals:[22]

- Restore the financial health and break free from the bankers' control
- Expand the federally funded work relief program for the unemployed
- End corruption in government and racketeering in key sectors of the economy
- Replace patronage with a merit-based civil service, with high prestige
- Modernize the infrastructure, especially transportation and parks

He achieved most of the first four goals in his first hundred days, as FDR gave him 20% of the entire national CWA budget for work relief. LaGuardia then collaborated closely with Robert Moses, with support from the governor, Democrat Herbert Lehman, to upgrade the decaying infrastructure. The city was favored by the New Deal in terms of funding for public works projects.

Ethnic politics

LaGuardia governed in an uneasy alliance with New York's Jews and liberal WASPs, together with Italian and German ethnics.[23]

LaGuardia was not an orthodox Republican. He also ran as the nominee of the American Labor Party, a union-dominated anti-Tammany left-wing group that supported Franklin D. Roosevelt for president beginning in 1936. LaGuardia supported Roosevelt, chairing the Committee of Independent Voters for Roosevelt and Wallace with Senator George Norris during the 1940 presidential election.

LaGuardia was the city's first Italian-American mayor, but was not a typical Italian New Yorker. He was a Republican Episcopalian who had grown up in Arizona, and had a Triestine Jewish mother[6] and a Catholic-turned-atheist Italian father. He reportedly spoke several languages, including Hebrew, Croatian, German, Italian, and Yiddish. LaGuardia was also a very active Freemason.

Fiorello LaGuardia statue at LaGuardia Place in Greenwich Village, NYC

Crime

LaGuardia loathed the gangsters who brought a negative stereotype and shame to the Italian community.[24] His first action as mayor was to order the chief of police to arrest mob boss Lucky Luciano on whatever charges could be found. LaGuardia then went after the gangsters with a vengeance, stating in a radio address to the people of New York in his high-pitched, squeaky voice, "Let's drive the bums out of town." In 1934, LaGuardia went on a search-and-destroy mission looking for mob boss Frank Costello's slot machines, which La Guardia executed with gusto, rounding up thousands of the "one armed bandits", swinging a sledgehammer and dumping them off a barge into the water for the newspapers and media. In 1935, La Guardia appeared at The Bronx Terminal Market to institute a city-wide ban on the sale, display, and possession of artichokes, whose prices were inflated by mobs. When prices went down, the ban was lifted.[25] In 1936, LaGuardia had special prosecutor Thomas E. Dewey, a future Republican presidential candidate, single out Lucky Luciano for prosecution. Dewey led a successful investigation into Luciano's lucrative prostitution operation, eventually sending Luciano to jail with a 30–50 year sentence. The case was made into the 1937 movie *Marked Woman*, starring Bette Davis.

LaGuardia proved successful in shutting down the burlesque theaters, whose naughty shows offended his puritanical sensibilities.[26]

Public works

LaGuardia's admirers credit him for, among other things, restoring the economic lifeblood of New York City during and after the Great Depression. He is given credit for many massive public works programs administered by his powerful Parks Commissioner Robert Moses and employed thousands of voters. The mayor's relentless lobbying for federal funds allowed New York to develop its economic infrastructure.

To obtain large-scale federal money the mayor became a close partner of Roosevelt and New Deal agencies such as CWA, PWA and WPA, which poured $1.1 billion into the city from 1934–39 . In turn he gave FDR a showcase for New Deal achievement, helped defeat FDR's political enemies in Tammany Hall (the Democratic party machine in Manhattan). He and Moses built highways, bridges and tunnels, transforming the physical landscape of New York City. The West Side Highway, East River Drive, Brooklyn Battery Tunnel, Triborough Bridge, and two airports (Floyd Bennett Field, and, later, LaGuardia Airport) were built during his mayoralty.

1939

1939 was a busy year, as he opened the 1939 New York World's Fair at Flushing Meadows-Corona Park, Queens, opened New York Municipal Airport No. 2 in Queens (later renamed Fiorello H. LaGuardia Field), and had the city buy out the Interborough Rapid Transit Company, thus completing the public takeover of the subway system. When the newspapers went on strike he read the funny papers on the radio.

Reform

Responding to popular disdain for the sometimes corrupt City Council, LaGuardia successfully proposed a reformed 1938 City Charter that created a powerful new New York City Board of Estimate, similar to a corporate board of directors.

Germany

He was an outspoken and early critic of Adolf Hitler and the Nazi regime. In a public address in 1934, LaGuardia warned, "Part of Hitler's program is the complete annihilation of the Jews in Germany." In 1937, speaking before the Women's Division of the American Jewish Congress, LaGuardia called for the creation of a special pavilion at the upcoming New York World's Fair, "a chamber of horrors" for "that brown-shirted fanatic".[27]

Gemma LaGuardia Gluck

LaGuardia's sister, Gemma LaGuardia Gluck (1881–1962),[28] and brother-in-law, Herman Gluck (a Hungarian Jew whom she met while teaching English in Europe), were living in Hungary and were arrested by the Gestapo on June 7, 1944,[29] when the Nazis took control of Budapest. Adolf Eichmann and Heinrich Himmler knew Gemma was Fiorello's sister and ordered her held as a political prisoner. She and Herman were deported to Mauthausen concentration camp in Austria, where he died, as Gemma learned from reading a newspaper account a year following her release.[30] [31] She was transferred from Mauthausen to the notorious women's concentration camp at Ravensbruck, located 50 miles from Berlin, where unbeknownst to Gemma at the time, her daughter Yolanda[28] (whose husband also died in the camps[32]) and baby grandson were also held for a year in a separate barrack.[33] Gemma, who was held in Block II of the camp and assigned prisoner #44139,[29] was one of the few survivors of this camp[34] and wrote about her time at Ravensbrück with great clarity.[35] [36] She also wrote that the Soviets were "violating girls and women of all ages", and about her, her daughter's and grandson's suffering as displaced persons in postwar Berlin, where the Germans abandoned them for a possible hostage exchange in April 1945, as the Russians were advancing. Gemma and her family did not speak German, and had no identity papers, money, or means of documenting where they had been. Gemma finally managed to get word to the Americans who contacted Fiorello, who had no idea where they were. He worked to get them on the immigration lists, but asserted in a letter, included in the appendix of Gemma's memoir, that her "case was the same as that of hundreds of thousands of displaced people" and "no exceptions can be made". Thus, despite Gemma's intimate connection with a powerful American politician, who was then director of the United Nations Relief and Rehabilitation Administration (UNRRA), it took two years to be cleared and sent to the United States. She returned to New York in May 1947, where she reunited with Fiorello four months before he died. As he had made no provision for her, she lived in very reduced circumstances, in a LaGuardia public housing project in Queens, New York, until her death in 1962.[28] [37] Gluck is believed to be the only American-born woman interned by the Nazis.

Courts

According to *Try and Stop Me* by Bennett Cerf, LaGuardia often officiated in municipal court. He handled routine misdemeanor cases, including, as Cerf wrote, a woman who had stolen a loaf of bread for her starving family. LaGuardia insisted on levying the fine of ten dollars. Then he said "I'm fining everyone in this courtroom fifty cents for living in a city where a person has to steal bread in order to eat!" He passed a hat and gave the fines to the defendant, who left the court with $47.50.[38] There is, however, no convincing proof of this anecdote.[39]

World War II

In 1941, during the run-up to American involvement in World War II, President Roosevelt appointed LaGuardia first director of the new Office of Civilian Defense (OCD). Roosevelt was an admirer of LaGuardia; after meeting Winston Churchill for the first time he described him as "an English Mayor LaGuardia."[40] The OCD was the national agency responsible for preparing for blackouts, air raid wardens, sirens, and shelters in case of German air raids. The government knew that such air raids were impossible but the goal was to psychologically mobilize many thousands of middle class volunteers to make them feel part of the war effort. LaGuardia remained Mayor of New York, shuttling back and forth with three days in Washington and four in the city in an effort to do justice to two herculean jobs. On top of this, he still performed other gestures, such as arranging police protection with his personal assurances for local artists Joe Simon and Jack Kirby, when they were threatened by Nazi supporters for their new patriotic comic book superhero, Captain America.[41] After the attack on Pearl Harbor in December 1941, his role was turned over to full-time director of OCD, James M. Landis. LaGuardia's popularity slipped away and he ran so poorly in straw polls in 1945 that he did not run for a fourth term.[42]

Unemployment ended and the city was the gateway for military supplies and soldiers sent to Europe, with the Brooklyn Navy Yard providing many of the warships and the garment trade provided uniforms. The city's great financiers, however, were less important in decision making than policy makers in Washington, and very high wartime taxes were not offset by heavy war spending. New York was not a center of heavy industry and did not see a wartime boom as defense plants were built elsewhere.[43]

FDR refused to make LaGuardia a general and was unable to provide fresh money for the city. By 1944, LaGuardia was frantically juggling the books to pay the city's bills. His successors realized that New York City could not support his fabulous infrastructure and high wages and pensions for teachers, police and city workers without borrowing more and more until it faced bankruptcy, which came in 1975.[44]

Later life and death

LaGuardia was the director general for the United Nations Relief and Rehabilitation Administration (UNRRA) in 1946.

The grave of Fiorello LaGuardia

A man of short stature, LaGuardia's height is sometimes given as 5 feet 0 inches (m). According to an article in *The New York Times,* however, his actual height was 5 feet 2 inches (m).[45]

He became a member of Phi Mu Alpha Sinfonia music fraternity.

He died of pancreatic cancer in his home at 5020 Goodridge Avenue, in the Riverdale section of the Bronx[46] at the age of 64 and is interred at Woodlawn Cemetery in the Bronx.[47]

Legacy

Historians have recognized La Guardia as among the best mayors in New York City history and perhaps among the greatest in modern U.S. history.[48]

Memorials

Awards and honors

- The United States Postal Service honored him with a 14¢ postage stamp in 1972.
- In 1940, La Guardia received The Hundred Year Association of New York's Gold Medal Award "in recognition of outstanding contributions to the City of New York".

The footstone of Fiorello LaGuardia

Photographs

- He got passed the laws to ban pinball machines as gambling devices around 1939, and many photos exist of him destroying the games prohibition-style, with sledge hammers or in large fires. The laws stayed on the books until 1976.

Landmarks and memorials

In New York

- La Guardia loved music, and was famous for spontaneously conducting professional and student orchestras. He once said that the "most hopeful accomplishment" of his administration as mayor was the creation of the High School of Music & Art in 1936, now the Fiorello H. LaGuardia High School of Music & Art and Performing Arts.[49]

- In addition to LaGuardia High School, a number of other institutions are also named for him, including LaGuardia Community College.
- LaGuardia Airport, the smallest of New York's three major currently operating airports, bears his name; the airport was voted the "greatest airport in the world" by the worldwide aviation community in 1960. La Guardia ordered construction of the airport after his TWA flight arrived at Newark, New Jersey. The arrival city for his airline ticket read "New York"; landing, instead, in the neighboring state outraged him and caused him to order the plane to fly to Brooklyn's Floyd Bennett Field. Not long afterward, the city voted to build a new airport in La Guardia's name.
- LaGuardia Place, a street in Greenwich Village which runs from Houston Street to Washington Square, is named for La Guardia; there is also a statue of the mayor on that street.
- On Staten Island, Masonic lodge #1130 at 236 Main Street is named after him.

In other states

- La Guardia Bridge in Prescott, Arizona on North Montezuma Avenue, is named after him.

Overseas

- "Ulica Fiorella La Guardije" (Fiorello La Guardia Street) is the name of a street in Rijeka, Croatia. La Guardia served in the U.S. consulate in Rijeka during the period before World War I when the city was under Austro-Hungarian rule and was known under its Italian name Fiume.
- *Rehov LaGuardia* (LaGuardia Street) is a major road and the name of a highway interchange on the Ayalon Highway in Tel Aviv.

In popular culture

In film

- In *The Court Martial of Billy Mitchell*, he is portrayed by Phil Arnold.
- He is mentioned in *Ghostbusters II*. During the supernatural chaos towards the end of the film, the current mayor of New York states, "My wife and I sat in bed having a two-hour conversation with Fiorello La Guardia. And he's been dead for forty years!"
- He is seen in the documentary *Grass* (1999) making an anti-prohibition speech.

In literature

- In *The Plot Against America* by Philip Roth, he is depicted as one of the leaders of the opposition against president Charles Lindbergh.
- In the timeline of Robert Heinlein's utopian novel *For Us, the Living* – written in 1939 but only published posthumously in 2003 – LaGuardia is elected President in 1951 and serves two terms as a militant reforming president, effectively nationalizing the banking system and instituting a system of Social Credit.

Online

- The New York Public Radio WNYC Archives maintains a Twitter feed that features "tweets" from transcripts of LaGuardia's *Talk to the People* radio program at twitter.com [50].

Onstage

- LaGuardia was the subject of the Pulitzer Prize-winning, hit Broadway musical *Fiorello!*. He was portrayed by actor Tom Bosley. The show ran for two years and closed in 1961.

See also

- LaGuardia and Wagner Archives
- LaGuardia Commission, a study on the effects of marijuana in U.S. society
- List of mayors of New York City
- New York City mayoral elections for votes in 1929, 1933, 1937 and 1941.

References

Notes

[1] "The Green Book: Mayors of the City of New York" (http://www.nyc.gov/html/dcas/html/features/greenbook_mayors.shtml) on the official NYC website.

[2] He wrote his name as a single word with no space between the "La" and the capitalized "G" which follows.

[3] He was ranked first in Melvin G. Holli, *The American Mayor* (1993)

[4] Kessner, Thomas (1989). *Fiorello H. LaGuardia and the Making of Modern New York.*

[5] Roberts, Sam (April 18, 2008). "The Giuliani Years: History; LaGuardia's Legacy Is Formidable, But It May Be Surpassed" (http://query.nytimes.com/gst/fullpage.html?res=9C01E2D81730F932A05751C1A9679C8B63). *The New York Times.*.

[6] Gross, Daniela. "Le radici triestine di Fiorello LaGuardia leggendario sindaco di New York City" (http://ilpiccolo.gelocal.it/dettaglio/le-radici-triestine-di-fiorello-laguardia-leggendario-sindaco-di-new-york-city/1313385) (in Italian). *Newspaper article.* Il Piccolo. pp. 1.. Retrieved April 23, 2010.

[7] L000007 LA GUARDIA, Fiorello Henry (http://bioguide.congress.gov/scripts/biodisplay.pl?index=) at the *Biographical Directory of the United States Congress* from the Biographical Directory of the United States Congress

[8] For one biographical account about Achille LaGuardia, Foraker, Sheila. "Achille La Guardia: Bandmaster of the 11th U.S. Infantry Territorial Brass" (http://www.territorialbrass.com/LaGuardia.html).. Retrieved December 1, 2010.

[9] "LaGuardia and Wagner Archives Photo Collection – Family Album # 2" (http://istrianet.org/istria/illustri/non-istrian/la_guardia/family-album2.htm).. Retrieved December 1, 2010.

[10] Zinn, Howard (2010). *LaGuardia in Congress* (http://books.google.com/books?id=2hNdePcjrdUC&pg=PA7). Cornell University Press. pp. 7. ISBN 978-0-8014-7617-4.. Retrieved November 13, 2011.

[11] Zinn, *LaGuardia in Congress* (1959)

[12] "R.L. Moran Led City Alderman". *The New York Times*. August 19, 1954.

[13] "Major Kelly Killed by His Own Pistol". *The New York Times*. July 23, 1930.

[14] "This Election Near A Collapse for Tammany," *The New York Times,* November 6, 1919.

[15] Zinn, *LaGuardia in Congress* pp. 226–30

[16] Bernstein, Irving (1966). *The Lean Years: A History of the American Worker, 1920–1933.* pp. 406–9.

[17] Zinn, *LaGuardia in Congress* pp. 267–70

[18] Zinn, *La Guardia in Congress,* (1959)

[19] Joseph McGoldrick, "The New York City Election of 1929," *The American Political Science Review,* Vol. 24, No. 3 (Aug., 1930), pp. 688–690 in JSTOR (http://www.jstor.org/stable/1946937)

[20] Stolberg, Mary M. *Fighting Organised Crime: Politics, Justice, and the Legacy of Thomas E. Dewey* (1995) pg. 229

[21] Arthur H. Mann, *LaGuardia Comes to Power 1933* (1969)

[22] Thomas Kessner, *Fiorello H. LaGuardia and the Making of Modern New York* (1989) ch 8–9

[23] Ronald H. Bayor, *Fiorello La Guardia: Ethnicity and Reform* (1993)

[24] Thomas Kessner, *Fiorello H. LaGuardia and the Making of Modern New York* (1989) pp 350–68

[25] Gray, Christopher (May 8, 1994). "Streetscapes/Bronx Terminal Market; Trying to Duplicate the Little Flower's Success" (http://www.nytimes.com/1994/05/08/realestate/streetscapes-bronx-terminal-market-trying-duplicate-little-flower-s-success.html). *The New York Times.*. Retrieved July 6, 2011.

[26] Friedman, Andrea (October 1996). "'The Habitats of Sex-Crazed Perverts': Campaigns against Burlesque in Depression-Era New York City" (http://www.jstor.org/stable/3704140). *Journal of the History of Sexuality* Vol. 7, No. 2, pp. 203–238.

[27] David M. Esposito, and Jackie R. Esposito, "La Guardia and the Nazis, 1933–1938." *American Jewish History* 1988 78(1): 38–53. ISSN: 0164-0178; quote from H. Paul Jeffers, *The Napoleon of New York* (2002) p. 233.

[28] Robert Leiter, *Jewish Exponent* Staff (January 31, 2008). "Pain & Triumph: The journey of a New York mayor's family".

[29] Nanda Herbermann, Hester Baer, Elizabeth Roberts Baer. *The Blessed Abyss: Inmate #6582 in Ravensbrück Concentration Camp for Women* (http://books.google.com/books?id=3JM3AdnmE18C&pg=PA257&lpg=PA257&dq=1961+gemma+gluck+my+story&source=bl&ots=ApQ_vZwimh&sig=JOLGQIBoS5ZR_GS2G2hO8eMKfSY&hl=en&sa=X&ei=b6dlUMbpFafG0AGN4oHoDA&ved=0CDUQ6AEwAQ#v=onepage&q=1961 gemma gluck my story&f=false). . Retrieved September 28, 2012.

[30] By Jon Kalish (Published June 26, 2007). "Ravensbruck's Famous Survivor: Memoir" (http://forward.com/articles/11031/ravensbruck-s-famous-survivor-/#). *The Jewish Daily Forward* (June 29, 2007). . Retrieved September 28, 2012.

[31] Donald Reid (May-August 2008). "America so far from Ravensbrück" (https://docs.google.com/viewer?a=v&q=cache:lFM8YhJcEswJ:www.histoire-politique.fr/documents/05/dossier/pdf/HP5-Donald-Reid-pdf.pdf+download+gemma+gluck+1961+ravensbruck&hl=en&gl=us&pid=bl&srcid=ADGEEShdg2sxeS_NSFLLSRhpisYVWrQRyVzTmgjQkU5oSwQJk95m-4rPlP-iR7g38eTt6fvznYiwLCED7la0zbDlCGh1Bo5HQd2Eclg-tEq9a40
sig=AHIEtbS_Uw7gUKHQ3iCa8qUQ1tPCTYxNrg). *Histoire@Politique: Politique, culture, société, N°5*. . Retrieved September 26, 2012.

[32] Gemma LaGuardia Gluck (1961). My Life.

[33] Mulligan, Katherine. "New book Reveals Holocaust Plight of La Guardia's Sister" (http://www.jewishrockland.org/page.aspx?id=162219). Jewish Federation of Rockland County. . Retrieved September 26, 2012.

[34] "Adolf Eichmann's List" (http://www.timesonline.co.uk/tol/news/world/europe/article3537593.ece). Times Online. .

[35] Sandee Brawarsky - *Jewish Week* Book Critic (April 13, 2007). "Mayor LaGuardia's Sister" (http://www.rememberwomen.org/Library/BkReviews/main.html). *The Jewish Week*. . Retrieved September 29, 2012.

[36] Gemma La Guardia Gluck (Author), Rochelle G. Saidel (Editor) (2007, originally published as *My Story* in 1961). *Fiorello's Sister: Gemma La Guardia Gluck's Story (Religion, Theology, and the Holocaust)*. Syracuse University Press.

[37] "Book reviews at Rememberwomen.org of *Fiorello's Sister: Gemma La Guardia Gluck's Story*, New Expanded Edition (2007) edited by Rochelle G. Saidel of *My Story by Gemma La Guardia Gluck* (1961), edited by S. L. Shneiderman" (http://www.rememberwomen.org/Library/BkReviews/main.html). . Retrieved September 29, 2012.

[38] Mikkelson, Barbara and David; (January 1, 2008). *LaGuardian Angel.* (http://www.snopes.com/glurge/laguardia.asp) Snopes. Retrieved January 31, 2008.

[39] Mikkelson, Barbara and David; (August 19, 2009). *LaGuardian Angel.* (http://www.snopes.com/glurge/laguardia.asp) Snopes. Retrieved February 29, 2012.

[40] Thompson, Kenneth W. (1996). *Virginia Papers on the Presidency* (http://books.google.com/books?id=dbxWAAAAYAAJ&q="English+Mayor+LaGuardia."&dq="English+Mayor+LaGuardia."&hl=en&ei=al2_TpecJ4PV0QGLyvTcBA&sa=X&oi=book_result&ct=result&resnum=4&ved=0CEAQ6AEwAw). Lanham: University Press of America. p. 114. ISBN 978-0-7618-0545-8. . Retrieved November 13, 2011.

[41] Cronin, Brian (2009). *Was Superman a Spy?*. Plume: The Penguin Group. pp. 135–136.

[42] Erwin Hargrove, "The Dramas of Reform," in James D. Barber, ed. *Political Leadership in American Government* (1964), p. 94.)

[43] Karl Drew Hartzell, *The Empire State At War, World War II* (1949)

[44] Thomas Kessner, "Fiorello H. LaGuardia." *History Teacher* (1993) 26(2): 151–159

[45] Chan, Sewell (December 4, 2006). "The Empire Zone: The Mayor's Tall Tales" (http://empirezone.blogs.nytimes.com/2006/12/04/the-mayors-tall-tales/). *The New York Times*. . Retrieved August 14, 2008.

[46] Jackson, Nancy Beth. "If You're Thinking of Living In/Fieldston; A Leafy Enclave in the Hills of the Bronx" (http://query.nytimes.com/gst/fullpage.html?res=950CE7D91E3CF934A25751C0A9649C8B63&scp=1&sq="richard+simon"+riverdale&st=nyt). *The New York Times*. February 17, 2002. Retrieved May 3, 2008. "Fiorello H. La Guardia, a three-time mayor of New York, lived and died at 5020 Goodridge Avenue."

[47] Fiorello Henry LaGuardia (http://www.findagrave.com/cgi-bin/fg.cgi?page=gr&GRid=596) at *Find a Grave*

[48] He was first in *The American Mayor* (1993) by Melvin G. Holli.

[49] Steigman, Benjamin: *Accent on Talent – New York's High School of Music & Art* Wayne State University Press, 1984 ISBN 0-686-87975-9

Further reading

- Bayor, Ronald H. *Fiorello La Guardia: Ethnicity and Reform.* (1993). 213 pp.
- Brodsky, Alyn. *The Great Mayor: Fiorello La Guardia and the Making of the City of New York.* (2003). 530 pp. popular biography; excerpt and text search (http://www.amazon.com/dp/0312287372)
- Capeci, Dominic J. "From Different Liberal Perspectives: Fiorello H. La Guardia, Adam Clayton Powell, Jr., and Civil Rights in New York City, 1941–1943," *The Journal of Negro History,* Vol. 62, No. 2 (Apr., 1977), pp. 160–173 in JSTOR (http://www.jstor.org/stable/2717176)

- Caro, Robert. *The Power Broker: Robert Moses and the Fall of New York.* (1973) excerpt and text search (http://www.amazon.com/gp/reader/0394720245)
- Garrett, Charles. *The La Guardia Years: Machine and Reform Politics in New York City.* (1961).
- Hecksher, August III. *When La Guardia Was Mayor: New York's Legendary Years.* (1978).
- Jeffers, H. Paul. *The Napoleon of New York: Mayor Fiorello La Guardia* 392 pp. popular biography online edition (http://www.questia.com/read/106909939?title=The Napoleon of New York: Mayor Fiorello La Guardia); also excerpt and text search (http://www.amazon.com/dp/0471024651)
- Kaufman, Herbert. "Fiorello H. La Guardia, Political Maverick" *Political Science Quarterly* 1990 105(1): 113–122. Issn: 0032-3195 in Jstor (http://www.jstor.org/pss/2151228)
- Kessner, Thomas. "Fiorello H. LaGuardia." *History Teacher* 1993 26(2): 151–159. Issn: 0018-2745 in Jstor (http://www.jstor.org/stable/494812)
- Kessner, Thomas. *Fiorello H. LaGuardia and the Making of Modern New York* (1989) the most detailed standard scholarly biography
- LaGuardia, Fiorello H. *The Making of an Insurgent: An Autobiography.* (1948)
- La Guardia Gluck, Gemma: Fiorello's Sister: La Guardia's Gluck's story, originally published in 1961, reissued in 2007 with new material, edited by Rochelle Saidel. Syracuse University Press; Expanded edition.
- Mann, Arthur H. *La Guardia: A Fighter Against His Times 1882–1933* (1959)
- Mann, Arthur H. *La Guardia Comes to Power 1933* (1969)
- Zinn, Howard. *LaGuardia in Congress* (1959) online edition (http://www.questia.com/read/92000679?title=Laguardia in Congress)

External links

- Obituary, *New York Times,* September 21, 1947 *La Guardia Is Dead; City Pays Homage To 3-Time Mayor* (http://www.nytimes.com/learning/general/onthisday/bday/1211.html)
- La Guardia and Wagner Archives/Fiorello H. La Guardia Collection (http://www.laguardiawagnerarchive.lagcc.cuny.edu/COLLECTIONS.aspx?ViwType=1&ColID=1)
  - oral interviews from the La Guardia and Wagner Archives/Fiorello H. La Guardia Oral History database (http://www.laguardiawagnerarchive.lagcc.cuny.edu/Advanced_Search.aspx?ViwType=1&ColID=1)
- Tiziano Thomas Dossena, "Fiorello La Guardia" in *Bridge Apulia USA*, No.3 (Italy, 1998) (http://www.dossena.org/tiziano/laguardia.html)
- 1919 passport photo, Fiorello LaGuardia (http://www.flickr.com/photos/puzzlemaster/5429338693/in/photostream)
- WNYC Archives blogs featuring Mayor La Guardia (http://www.wnyc.org/tags/mayor_laguardia/)

Uta_Hagen

Uta Hagen	
 Hagen with Paul Robeson in the 1943–1945 Theatre Guild production of *Othello*	
Born	Uta Thyra Hagen12 June 1919Göttingen, Germany
Died	14 January 2004 (aged 84)New York City, New York, U.S.
Occupation	Actress
Years active	1937–2001
Spouse(s)	José Ferrer (m.1938-1948; divorced) Herbert Berghof (m.1957-1990; his death)

Uta Thyra Hagen (12 June 1919 – 14 January 2004) was a German-born American actress and drama teacher. She originated the role of Martha in the 1963 Broadway premiere of *Who's Afraid of Virginia Woolf?* by Edward Albee (who called her "a profoundly truthful actress"). Hagen was on the Hollywood blacklist, in part because of her association with Paul Robeson, and this curtailed film opportunities, focusing her to perform in New York theaters. She won the Tony Award three times. She later became a highly influential acting teacher at New York's Herbert Berghof Studio and authored best-selling acting texts, *Respect for Acting*, with Haskel Frankel,[1] and *A Challenge for the Actor*. She was elected to the American Theatre Hall of Fame in 1981.[2]

Life and career

Early life

Born in Göttingen, Germany, the daughter of Thyra A. (née Leisner) and Oskar Frank Leonard Hagen,[3] Hagen and her family emigrated to the United States in 1924, when her father received a position at Cornell University.[4] She was raised in Madison, Wisconsin. She appeared in productions of the University of Wisconsin High School and in summer stock productions of the Wisconsin Players. She studied acting briefly at the Royal Academy of Dramatic Art in 1936.[5] After spending one semester at the University of Wisconsin, where her father was the head of the department of art history, she left for New York City in 1937.[6] Her first professional role was as Ophelia opposite Eva Le Gallienne in the title role of *Hamlet* in Dennis, Massachusetts in 1937.

Career

Hagen was cast, early on, as Ophelia by the actress-manager Eva LeGallienne. From there, Hagen went on to play the leading ingenue role of Nina in a Broadway production of Anton Chekhov's *The Seagull* which featured Alfred Lunt and Lynn Fontanne. It was 1938; Hagen was just 18. This experience left an indelible mark on the young actress, as she later reflected, "My next job was Nina in The Seagull, [her Broadway bow] with the Lunts, on Broadway. That sounds incredible, too. They were an enormous influence on my life." She admired "their passion for the theatre, and their discipline. It was a 24-hour-a-day affair, and I never forgot it—never!"[7] The New York Times' critic Brooks Atkinson hailed her Nina as "grace and aspiration incarnate".[8]

She played George Bernard Shaw's Saint Joan (1951) on Broadway, and Desdemona in a production which toured and played Broadway, featuring Paul Robeson as Shakespeare's Othello and her then-husband Jose Ferrer as Iago. She took over the role of Blanche DuBois in *A Streetcar Named Desire* for the national tour, which was directed not by Elia Kazan who had directed the Broadway production but by Harold Clurman. Hagen had had a revelatory experience when she first worked with Clurman in 1947. In *Respect for Acting*, she credited her discoveries with Clurman as the springboard for what she would later explore with her husband Herbert Berghof: "how to find a true technique of acting, how to make a character flow through me". She played Blanche (on the road and on Broadway) opposite at least four different Stanley Kowalskis, including Anthony Quinn and Marlon Brando. Through interviews with her and contemporary criticism, the report is that Hagen's Blanche refocused the audience's sympathies with Blanche rather than with Stanley (where the Brando/Kazan production had leaned). Primarily noted for stage roles, Hagen won her first Tony Award in 1951 for her performance as the self-sacrificing wife Georgie in Clifford Odets' *The Country Girl*. She won again in 1963 for originating the role of the "I-wear-the-pants-in-this-family-because-somebody's-got-to" Martha in Edward Albee's *Who's Afraid of Virginia Woolf?*. (An original cast recording was made of this show.) In 1981 she was elected to the American Theatre Hall of Fame and in 1999 received a "Special Lifetime Achievement Tony Award".

Hagen considered the period between 1938 and 1947 as *the transitional years of my career, during which I lost my way and a love of acting until I finally regained it to begin a true life in the theater."*[8]

Although she appeared in some movies, because of the Hollywood blacklist she had more limited output in film and on television, not making her cinematic debut until 1972. She would later comment about being blacklisted, "that fact kept me pure".

Although Hagen played a character with a German accent in one of her best-known Hollywood films, *The Boys from Brazil*, in which her scene is with Laurence Olivier, and a Russian accent in *The Other*, she had simply assumed the accent for those roles (Hagen was raised in Wisconsin). She was nominated for a Daytime Emmy Award as "Outstanding Supporting Actress in a Drama Series" for her performance on the television soap opera *One Life to Live*.

She taught at HB Studio, a well-known New York City acting school on a cobblestone, tree-shaded street in the West Village. She began there in 1947, and married its co-founder, Herbert Berghof, on January 25, 1957. Later in her life, Hagen undertook a return to the stage, earning accolades for leading roles in *Mrs. Warren's Profession* (1985), *Collected Stories*, and *Mrs. Klein*. After Berghof's death in 1990 she became the school's chairperson.

Hagen was an influential acting teacher who taught, among others, Matthew Broderick, Christine Lahti, Amanda Peet, Jason Robards, Sigourney Weaver, Katie Finneran, Liza Minnelli, Whoopi Goldberg, Jack Lemmon, Charles Nelson Reilly, Manu Tupou, Debbie Allen, Herschel Savage, Jon Stewart, and Al Pacino. She was a voice coach to Judy Garland, teaching a German accent, for the picture *Judgment at Nuremberg*. Garland's performance earned her an Academy Award nomination.

She also wrote *Respect for Acting* (1973) and *A Challenge for the Actor* (1991), which advocate realistic acting (as opposed to pre-determined "formalistic" acting). In her mode of realism, the actor puts his own psyche to use in finding identification with the role," trusting that a form will result.[9] In *Respect for Acting*, Hagen credited director Harold Clurman with a turn-around in her perspective on acting:

> *"In 1947, I worked in a play under the direction of Harold Clurman. He opened a new world in the professional theatre for me. He took away my 'tricks'. He imposed no line readings, no gestures, no positions on the actors. At first I floundered badly because for many years I had become accustomed to using specific outer directions as the material from which to construct the mask for my character, the mask behind which I would hide throughout the performance. Mr Clurman refused to accept a mask. He demanded ME in the role. My love of acting was slowly reawakened as I began to deal with a strange new technique of evolving in the character. I was not allowed to begin with, or concern myself at any time with, a preconceived form. I was assured that a form would result from the work we were doing."*

Hagen later "disassociated" herself from her first book, *Respect for Acting*.[10] In *Challenge for the Actor* she redefined a term which she had initially called "substitution", an esoteric technique for alchemizing elements of an actor's life with his/her character work, calling it "transference" instead. Though Hagen wrote that the actor should identify the character they play with feelings and circumstances from their (the actor's) own life, she also makes clear that

> *"Thoughts and feelings are suspended in a vacuum unless they instigate and feed the selected actions, and it is the characters' actions which reveal the character in the play."*

Respect for Acting is used as a textbook for many college acting classes. She also wrote a 1976 cookbook, *Love for Cooking*. In 2002, she was awarded the National Medal of the Arts by President George W. Bush at a ceremony held at the White House.

In 2001, Hagen released a video entitled *Uta Hagen's Acting Class*, a two-part set that captures her master classes.[11]

Harvey Korman talks about studying under her during his Archive of American Television interview in 2004.[12] David Hyde Pierce worked with Hagen in the Richard Alfieri play "Six dance lessons in six weeks", at the Geffen Playhouse in 2001.[13] Hyde Pierce spoke at her 2004 memorial at Manhattan's Majestic Theater.[14]

Work

Stage

- *The Seagull* - 1938
- *The Happiest Days* - 1939
- *Key Largo* - 1939
- *Vickie* - 1942
- *Othello* - 1943
- *The Whole World Over* - 1947
- *A Streetcar Named Desire* - 1947
- *The Country Girl* - 1950
- *Saint Joan* - 1951
- *In Any Language* - 1952
- *The Magic and The Loss* - 1954
- *Island of Goats* - 1955
- *Who's Afraid of Virginia Woolf?* - 1962
- *The Cherry Orchard* - 1968
- *You Never Can Tell* - 1986
- *Charlotte* - 1980
- *Mrs. Klein* - 1995
- *Collected Stories* - 1998

Film

- *The Other* - 1972
- *The Boys from Brazil* - 1978
- *A Doctor's Story* - 1984
- *Reversal of Fortune* - 1990
- *Limón: A Life Beyond Words* - 2001

Television

- *Victory* - 1945
- *A Month in the Country* - 1959
- *The Day Before Sunday* - 1970
- *The Twilight Zone (1985 TV series) - 1986*
- *Seasonal Differences* - 1987
- *The Sunset Gang* - 1991
- *King of the Hill* - 1999
- *Oz* - 1999

Awards and Nominations

- 1951 Tony Award, Actress- Play, *The Country Girl"*[15]
- 1963 Tony Award, Actress- Play, " Who's Afraid of Virginia Woolf?"
- Special 1999 Tony Award for Lifetime Achievement[16]
- 2002 National Medal of Arts

Interestingly, Hagen was quoted, saying, "Awards don't really mean much."[17]

Quotes

- "Once in a while, there's stuff that makes me say, That's what theatre's about. It has to be a human event on the stage, and that doesn't happen very often."[18]

See also

- Stanislavski
- Mikhail Chekhov
- Lee Strasberg
- Stella Adler
- Michelle Danner

References

[1] Hagen, Uta (1973). *Respect for Acting*. New York: Wiley Publsihing, Inc..
[2] *The New York Times*, March 3, 1981 - *26 Elected to the Theater Hall of Fame* (http://www.nytimes.com/1981/03/03/theater/26-elected-theater-hall-fame-26-broadway-voted-into-theater-hall-fame.html)
[3] (http://books.google.ca/books?id=up4YAAAAIAAJ&q=uta+hagen+gottingen&dq=uta+hagen+gottingen&hl=en&redir_esc=y)
[4] Port of New York, passenger list of the S.S. *Luetzow*, September 4, 1924, sheet 41.
[5] Port of New York, passenger list of the S.S. *Westernland*, December 24, 1936, sheet 165.
[6] "Lady Invincible" (http://digicoll.library.wisc.edu/cgi-bin/wiacrev/wiacrev-idx?type=HTML&rgn=DIV1&byte=678186&q1=&q2=&q3=), *Wisconsin Academy Review*, vol. 46, issue 4, Fall 2000.
[7] http://www.playbill.com/news/article/83844-STAGE-TO-SCREENS-A-Chat-with-Theresa-Rebeck-Remembering-Uta-Hagen/all
[8] Gussow, Mel (15 January 2004). "Uta Hagen, Tony-Winning Broadway Star and Teacher of Actors, Dies at 84" (http://query.nytimes.com/gst/fullpage.html?res=9804E5D61030F936A25752C0A9629C8B63&pagewanted=2). *The New York Times*. .
[9] Hagen, Uta 1991. *A Challenge for the Actor*. New York: Scribner's. ISBN 0-684-19040-0
[10] [Interview in Playbill at www.playbill.com]
[11] "Uta Hagen, " Internet Movie Databank,http://us.imdb.com
[12] Archive of American Television - Harvey Korman, 20 April 2004 (http://www.emmytvlegends.org/interviews/people/harvey-korman) www.emmytvlegends.org (accessed 18 June 2012)
[13] Richard Alfieri's "Six dance lessons in six weeks" (http://www.sixdancelessonsinsixweeks.com/photo-gallery/) Photo gallery (accessed 19 June 2012)
[14] Uta Hagen Memorial, New York Times, 20 March 2004 (http://www.nytimes.com/2004/03/20/arts/uta-hagen-memorial.html?ref=utahagen) (accessed 18 June 2012)
[15] http://www.infoplease.com/ipa/A0153406.html
[16] http://www.hbstudio.org/classes/meetuta.htm
[17] http://www.brainyquote.com/quotes/authors/u/uta_hagen.html
[18] http://www.brainyquote.com/quotes/authors/u/uta_hagen.html#ixzz1I6ezPnCJ

External links

- Uta Hagen (http://www.ibdb.com/person.asp?ID=43654) at the Internet Broadway Database
- Uta Hagen (http://www.lortel.org/LLA_archive/index.cfm?search_by=people&first=Uta&last=Hagen&middle=) at the Internet Off-Broadway Database
- Uta Hagen (http://www.imdb.com/name/nm353467/) at the Internet Movie Database
- Uta Hagen and Herbert Berghof papers, 1889-2004 and undated (http://catalog.nypl.org/record=b16890590~S1) held by the Billy Rose Theatre Division, New York Public Library for the Performing Arts

Winged Victory (play)

WARNING: Article could not be rendered - ouputting plain text.
Potential causes of the problem are: (a) a bug in the pdf-writer software (b) problematic Mediawiki markup (c) table is too wide

Winged Victory is a play and, later, a Winged Victory (film)film by Moss Hart, originally created and produced by the U.S. Army Air Forces during World War II as a morale booster and as a fundraiser for the Army Emergency Relief Fund. Upon recommendation of Lt. Col. Dudley DeanDudley S. Dean, who had been approached with the idea by talent agent Irving Paul LazarIrving Lazar and Lt. Benjamin Landis, Air Forces Commanding General Henry H. Arnold commissioned famed playwright Hart to create a stage drama that would depict both the training and work of airmen and also the commitment and devotion to duty with which they carried out this work. Arnold gave Hart complete access to Air Forces resources.Hart traveled the United States visiting Air Forces bases and seeing firsthand how training was carried out. He then wrote an extremely patriotic and uplifting stage play based on his research. Hart partnered with Irving Paul LazarIrving Lazar to produce the play and to gather actors for the huge (nearly 300) cast. Casting calls went out not only to well-known names but to Army units around the country for Air Forces personnel with theatrical or performing background. All male personnel involved with the show were service members of the U.S. Army Air Forces and were billed with their actual ranks. Among the then-notable or eventually-to-become-notable members of the cast and crew were Keith Andes, Alan Baxter (actor)Alan Baxter, Don Beddoe, Whit Bissell, Sascha Brastoff, Red Buttons, Lee J. Cobb, Mario Lanza, Mark Daniels (actor)Mark Daniels, Leonard De Paur, Brad Dexter, John Forsythe, Peter Lind Hayes, Harry Horner, Richard Travis (actor)Richard Travis, Karl Malden, Billy and Bobby Mauch, Kevin McCarthy (actor)Kevin McCarthy, Gary Merrill, Ray Middleton, Barry Nelson, Edmond O'Brien, Walter Reed (actor)Walter Reed, George Reeves, Martin Ritt, Archie Robbins, David Rose, Henry Rowland (actor)Henry Rowland, Alfred Ryder, Howard Shoup, Henry SlateHenry and Jack Slate, Claude Stroud, Don Taylor (actor)Don Taylor, and Victor Sen Yung. Most of the relatively few roles for women in the play were portrayed by wives of the actors. Winged Victory tells the story of a group of recruits struggling to make it through pilot training. The trainees are a cross-section of American young men. Their personal lives, their families and sweethearts make up a small part of the story, but most of the drama focuses on training and camaraderie. Music plays a large part in the play, and most of the huge cast were primarily members of a chorus under the direction of famed choral leader Leonard de Paur. Among the musical numbers were "My Little Dream Book of Memories," and the stirring title anthem, "Winged Victory".The play opened in Boston, Massachusetts in the fall of 1943, in a pre-Broadway theatreBroadway run, and was a huge success. It then opened in New York at the Forty-Fourth Street Theatre on 20 November 1943 and became a smash hit, playing to over 350,000 people in 226 performances. During the New York run, performers in the show also toured local military camps, providing entertainment to the troops. A number of Winged Victory actors also rehearsed and produced a production of the play Yellow Jack (play and film)Yellow Jack by Sidney Kingsley, under the direction of Martin Ritt, which played briefly and simultaneously on Broadway. The successful run of Winged Victory on Broadway ended only in order that the entire cast travel to Hollywood to do the Winged Victory (film)film version. Twentieth Century Fox had purchased the rights and contracted the full cast and in the summer of 1944 produced a film version (which featured a few actors who had not been in the play, including Lon McCallister and Judy Holliday), under the direction of George Cukor.Following filming, the company embarked on a national tour of the play, performing 445 times for over 800,000 people in Los Angeles, San Francisco, Denver, Kansas City, MissouriKansas City, Chicago, Detroit, Baltimore (where the theatre proved too small to admit the huge section of a bomber, part of the set for one scene), Washington D.C., and Richmond, VirginiaRichmond. At tour's end in April, 1945, the cast and crew were dispersed throughout the Army Air Forces, many of them transferring to the First Motion Picture Unit in California, there to make training films. The Winged Victory company (officially known as the 31st AAF Base Unit) officially disbanded in November, 1945.Unit History, 31st AAF Base Unit, Air Force Historical Research Agency, Maxwell Air Force BaseWertheim, Albert: "The Dramatic Art of Uncle Sam: The Government, Drama, and World War II." American Drama, Winter 2004Due both to its enormous cast and staging demands as well as to the extremely era-specific nature of the play, Winged Victory is one of the biggest hits in Broadway history never to have a second production anywhere.References

Mario_Lanza

Mario Lanza (January 31, 1921 – October 7, 1959), was an American tenor, singer, actor, and Hollywood movie star of the late 1940s and the 1950s. The son of Italian emigrants, he began studying to be a professional singer at the age of 16. He was born Alfredo Arnold Cocozza. He changed his name when he signed with Metro-Goldwyn-Mayer. His mother's maiden name was Maria Lanza.

MGM still, circa 1949

After appearing at the Hollywood Bowl in 1947, Lanza signed a seven-year contract with MGM's head, Louis B. Mayer, who saw his performance and was impressed by his singing. Prior to this, the adult Lanza had sung only two performances of an opera. The following year (1948), however, he would sing the role of Pinkerton in Puccini's *Madama Butterfly* in New Orleans.

His movie debut was in *That Midnight Kiss* (1949) with Kathryn Grayson and Ethel Barrymore. The following year, in *The Toast of New Orleans*, his featured popular song "Be My Love" became his first million-selling hit. In 1951, he played the role of Enrico Caruso (1873–1921), his tenor idol, in the biopic, *The Great Caruso*, which produced another million-seller with "The Loveliest Night of the Year" (a song which used the melody of *Sobre las Olas*). *The Great Caruso* was the top-grossing film that year.[1]

The title song of his next film, *Because You're Mine*, was his final million-selling hit song. The song went on to receive an Academy Award nomination for Best Original Song. After recording the soundtrack for his next film, *The Student Prince* he embarked upon a protracted battle with Studio Head Dore Schary arising from artistic differences with director Curtis Bernhardt, and was eventually dismissed by MGM.[2]

Lanza was known to be "rebellious, tough, and ambitious",[3] and during most of his film career, he suffered from addictions to overeating and alcohol which had a serious effect on his health and his relationships with directors, producers and, occasionally, other cast members. Hollywood columnist Hedda Hopper writes that "his smile, which was as big as his voice, was matched with the habits of a tiger cub, impossible to housebreak". She adds that he was the "last of the great romantic performers".[4] He made three more films before dying of a pulmonary embolism at the age of 38. At the time of his death in 1959 he was still "the most famous tenor in the world".[5] Author Eleonora Kimmel concludes that Lanza "blazed like a meteor whose light lasts a brief moment in time".[6]

Early years

Born **Alfredo Arnold Cocozza** in Philadelphia, Pennsylvania, he was exposed to classical singing at an early age by his Abruzzese-Molisan Italian parents, Maria (Lanza) and Antonio Cocozza.[7] [8] By the age of 16, his vocal talent had become apparent. Starting out in local operatic productions in Philadelphia for the YMCA Opera Company while still in his teens, he later came to the attention of longtime (1924-1949) principal Boston Symphony conductor Serge Koussevitzky, who in 1942 provided young Cocozza with a full student scholarship to the Berkshire Music Center at Tanglewood, Massachusetts. Reportedly, Koussevitzky would later tell him, "Yours is a voice such as is heard once in a hundred years."

Opera career

His opera debut, as Fenton in Otto Nicolai's *The Merry Wives of Windsor* (in English), came at the Berkshire Music Festival in Tanglewood on August 7, 1942, after a period of study with conductors Boris Goldovsky and Leonard Bernstein. This was when Cocozza adopted the stage name **Mario Lanza**, for its similarity to his mother's maiden name, Maria Lanza.

His performances at Tanglewood won him critical acclaim, with Noel Straus of *The New York Times* hailing the 21-year-old tenor as having "few equals among tenors of the day in terms of quality, warmth and power". Herbert Graf subsequently wrote in *Opera News* (October 5, 1942), "A real find of the season was Mario Lanza [...] He would have no difficulty one day being asked to join the Metropolitan Opera." Lanza sang Nicolai's Fenton twice at Tanglewood, in addition to appearing there in a one-off presentation of Act III of Puccini's *La bohème* with the noted Mexican soprano Irma González, baritone James Pease and mezzo-soprano Laura Castellano. Music critic Jay C. Rosenfeld wrote in *The New York Times* of August 9, 1942, "González as Mimì and Mario Lanza as Rodolfo were conspicuous by the beauty of their voices and the vividness of their characterizations." In an interview shortly before her own death in 2008, González recalled that Lanza was "very correct, likeable, with a powerful and beautiful voice".

Lanza as Giuseppe Verdi's Otello

His budding operatic career was interrupted by World War II, when he was assigned to Special Services in the U.S. Army Air Corps. He appeared in the wartime shows *On the Beam* and *Winged Victory*. He also appeared in the film version of the latter (albeit as an unrecognizable member of the chorus). He resumed his singing career with a concert in Atlantic City with the NBC Symphony Orchestra in September 1945 under Peter Herman Adler, subsequently his mentor. The following month, he replaced tenor Jan Peerce on the live CBS radio program *Great Moments in Music* on which he made six appearances in four months, singing extracts from various operas and other works.

He studied with Enrico Rosati for fifteen months, and then embarked on an 86-concert tour of the United States, Canada and Mexico between July 1947 and May 1948 with bass George London and soprano Frances Yeend. Reviewing his second appearance at Chicago's Grant Park in July 1947 in the *Chicago Sunday Tribune*, Claudia Cassidy praised Lanza's "superbly natural tenor" and observed that "though a multitude of fine points evade him, he possesses the things almost impossible to learn. He knows the accent that makes a lyric line reach its audience, and he knows why opera is music drama."

In April 1948, Lanza sang two performances as Pinkerton in Puccini's *Madama Butterfly* for the New Orleans Opera Association conducted by Walter Herbert with stage director Armando Agnini. Laurence Odel in the *St. Louis News*, Laurence Odel wrote, "Mario Lanza performed ... Lieutenant Pinkerton with considerable verve and dash. Rarely have we seen a more superbly romantic leading tenor. His exceptionally beautiful voice helps immeasurably." Following the success of these performances, he was invited to return to New Orleans in 1949 as Alfredo in Verdi's *La traviata*. But, as biographer Armando Cesari wrote, Lanza by 1949 "was already deeply engulfed in the Hollywood machinery and consequently never learned [that key mid-Verdi tenor] role."

Film career

A concert at the Hollywood Bowl in August 1947 had brought Lanza to the attention of Louis B. Mayer, who promptly signed Lanza to a seven-year film contract with Metro-Goldwyn-Mayer. This proved to be a turning point in the young singer's career. The contract required him to commit to the studio for six months, and at first Lanza believed he would be able to combine his film career with his operatic and concert one. In May 1949, he made his first commercial recordings with RCA Victor. His rendition of the aria "Che gelida manina" (from *La bohème*) from that session was subsequently awarded the prize of Operatic Recording of the Year by the (United States) National Record Critics Association.

The Toast of New Orleans

Lanza's first two starring films, *That Midnight Kiss* and *The Toast of New Orleans*, were commercial successes, and in 1950 his recording of "Be My Love" became the first of three million-selling singles for the young singer, earning him enormous fame in the process. While at MGM, Lanza worked closely with the Academy Award-winning conductor, composer, and arranger Johnny Green. In a 1977 interview with Lanza biographer Armando Cesari, Green recalled that the tenor was insecure about the manner in which he had become successful, and was keenly aware of the fact that he had become a Hollywood star before first having established himself on the operatic stage.

> "Had [Lanza] been already a leading tenor, if not *the* leading tenor at the Met[ropolitan Opera House], and come to Hollywood in between seasons to make a picture, he would have had [the security of

having] the Met as his home," Green remarked. According to Green, Lanza possessed "the voice of the next Caruso. [Lanza] had an unusual, very unusual quality...a tenor with a baritone color in the middle and lower registers, and a great feeling for the making of music. A great musicality. I found it fascinating, musically, to work with [him]".

The Great Caruso

In 1951, Lanza portrayed Enrico Caruso in *The Great Caruso*, which proved an astonishing success, though it did not adhere strictly to the facts of Caruso's life. At the same time, Lanza's increasing popularity exposed him to intense criticism by some music critics, including those who had praised his work years earlier. His performance earned him compliments from the subject's son, Enrico Caruso Jr., a tenor in his own right. Shortly before his own death in 1987, Enrico Jr. wrote in *Enrico Caruso: My Father and My Family* (posthumously published by Amadeus in 1990) that, "I can think of no other tenor, before or since Mario Lanza, who could have risen with comparable success to the challenge of playing Caruso in a screen biography... Lanza was born with one of the dozen or so great tenor voices of the century, with a natural voice placement, an unmistakable and very pleasing timbre, and a nearly infallible musical instinct." He went on to praise Lanza's tempi and phrasing, "flawless" diction, and "impassioned" delivery, adding that, "All are qualities that few singers are born with and others can never attain." In conclusion, he wrote that, "Lanza excelled in both the classical and the light popular repertory, an accomplishment that was beyond even my father's exceptional talents."

The Student Prince

In 1952, Lanza was dismissed by MGM after he had pre-recorded the songs for *The Student Prince*. The reason most frequently cited in the tabloid press at the time was that his recurring weight problem had made it impossible for him to fit into the costumes of the Prince. However, as his biographers Cesari and Mannering have established, Lanza was not overweight at the beginning of the production, and it was, in fact, a disagreement with director Curtis Bernhardt over Lanza's singing of one of the songs in the film that led to Lanza walking off the set. MGM refused to replace Bernhardt, and the film was subsequently made with actor Edmund Purdom miming to Lanza's voice. Ironically, the eventual director of the film was Richard Thorpe, the same man whom Lanza had pleaded with MGM to replace Bernhardt, and with whom the tenor had enjoyed an excellent working relationship in *The Great Caruso*.

Tenor Richard Tucker (left) speaking with Lanza in 1958 at Tucker's Covent Garden debut.

Depressed by his dismissal, and with his self-confidence severely undermined, Lanza became a virtual recluse for more than a year, frequently seeking refuge in alcoholic binges. During this period, Lanza also came very close to bankruptcy as a result of poor investment decisions by his former manager, and his lavish spending habits left him owing about $250,000 in back taxes to the IRS.

Serenade

Lanza returned to an active film career in 1955 in *Serenade,* released by Warner Bros. However the film was not as successful as his previous films, despite its strong musical content, including arias from *Der Rosenkavalier*, *Fedora*, *L'arlesiana*, and *Otello*, as well as the Act III duet from *Otello* with soprano Licia Albanese. He then moved to Rome, Italy in May 1957, where he worked on the film *Seven Hills of Rome*, and returned to live performing in a series of acclaimed concerts throughout the UK, Ireland and mainland Europe. Despite a number of cancellations, which resulted from his failing health during this period, Lanza continued to receive offers for operatic appearances, concerts, and films. In 1958, for example, he gave a concert tour of the UK.[9]

In September 1958, he made a number of operatic recordings at the Rome Opera House for the soundtrack of what would turn out to be his final film, *For the First Time*. The Artistic Director of the Rome Opera, Riccardo Vitale, offered him the role of Canio in *Pagliacci* in the theater's 1960/61 season. Lanza received offers from the management of the La Scala and San Carlo opera houses. However, his health continued to decline, with the tenor suffering from a variety of ailments, including phlebitis and acute high blood pressure. His old habits of overeating and crash dieting, coupled with binge drinking, compounded his problems. . On *The story of Variety with Michael Grade* (BBC 4, 30 May 2012) it was revealed that Lanza was often violent when drunk and that he suffered from the DT's, seeing things that were not there.

Death

In April 1959, Lanza suffered a minor heart attack followed in August by double pneumonia. He died in Rome in October at only 38 from a pulmonary embolism after undergoing a controversial weight loss program colloquially known as "the twilight sleep treatment," which required its patients to be kept immobile and sedated for prolonged periods. At his funeral were singers Maria Caniglia and Lidia Nerozzi and actors Franco Fabrizi and Enzo Fiermonte. American crooner Frank Sinatra sent his condolences by telegram.[10]

From the film Toast of New Orleans, as Lt. Pinkerton USN, in recreation of the opera Madama Butterfly, the only full professional opera in which Lanza ever appeared

Betty, his widow, returned to Hollywood with their four children but died five months later at the age of 37. Cesari cited the apparent cause of her death, according to the coroner, as "asphyxiation resulting from a respiratory ailment for which she had been receiving medication." In 1991, Marc, their youngest child, died of a heart attack at 37; six years later, Colleen, their elder daughter, was killed at 48 after being struck by two passing vehicles on a highway. Damon, the couple's oldest child and elder son, died in August 2008 of a heart attack at 55.

Legacy

Lanza's brief career covered opera, radio, concerts, recordings and motion pictures. He was the first RCA Victor Red Seal artist to win a gold disc and the first artist to sell 2 1/2 million albums. This highly influential performer has been credited with inspiring successive generations of other opera singers, including Plácido Domingo, Luciano Pavarotti, Leo Nucci and José Carréras. Singers with seemingly different backgrounds and influences were also inspired by him, even including his RCA Victor label-mate Elvis Presley.

Lanza was referred to by some sources as the "new Caruso" after his "instant success" in Hollywood films,[11] while MGM hoped he would become the movie studio's "singing Clark Gable" for his good looks and powerful voice.[3]

In 1994, outstanding tenor José Carréras paid tribute to Lanza during a worldwide concert tour, saying of him, "If I'm an opera singer, it's thanks to Mario Lanza."[12] His equally outstanding colleague Plácido Domingo echoed these comments in a 2009 CBS interview with, "Lanza's passion and the way his voice sounds are what made me sing opera. I actually owe my love for opera ... to a kid from Philadelphia."[13]

Even today "the magnitude of his contribution to popular music is still hotly debated," and because he appeared on the operatic stage only twice, many critics feel that he needed to have had more "operatic quality time" in major theaters before he could be considered a star of that art form.[5] His films, especially *The Great Caruso*, influenced numerous future opera stars, including Joseph Calléja,[14] José Carréras, Plácido Domingo and Luciano Pavarotti.[5] According to opera historian Clyde McCants, "Of all the Hollywood singers who performed operatic music . . . the one who made the greatest impact was Mario Lanza."[15] Hollywood gossip columnist Hedda Hopper concluded that "there had never been anyone like Mario, and I doubt whether we shall ever see his like again".[4]

Portrayal on stage

In October 2007, Charles Messina directed the big budget musical *Be My Love: The Mario Lanza Story*, written by Richard Vetere, about Lanza's life, which was produced by Sonny Grosso and Phil Ramone. It premiered at The Tilles Center for the Performing Arts in Greenvale, New York.[16]

Legacy

The persona and music of Mario Lanza are featured in the 1994 Peter Jackson film *Heavenly Creatures.*

Miljenko Jergovic mentions Lanza in his *Dvori od oraha* (*The Mansion in Walnut*) novel of 2003 as a part of story about Luka Sikiric.

Mario Lanza Boulevard is a roadway in the Eastwick section of Lanza's native Philadelphia, close to Philadelphia International Airport and ending on the grounds of the John Heinz National Wildlife Refuge.

The Mario Lanza Institute and Museum, which honors Lanza's legacy and also provides scholarships to young singers, is located at 712 Montrose Street in South Philadelphia. Link label [17]

Philadelphia's Queen Street Park was renamed for Lanza in 1967. Link label [18]

Lanza was born at 636 Christian Street in South Philadelphia. A Pennsylvania Historical Museum Commission marker stands outside of the house.

A large mural dedicated to Lanza stands at Broad and Reed Streets in Philadelphia.

In 1998, a Golden Palm Star on the Palm Springs, California, Walk of Stars was dedicated to him.[17]

Filmography

- *Winged Victory*, Twentieth Century-Fox 1944 (uncredited chorus member)
- *That Midnight Kiss*, MGM 1949
- *The Toast of New Orleans*, MGM 1950
- *The Great Caruso*, MGM 1951
- *Because You're Mine*, MGM 1952
- *The Student Prince*, MGM 1954 (voice only)
- *Serenade*, Warner Bros. 1956
- *Seven Hills of Rome*, MGM 1958
- *For the First Time*, MGM 1959

Select recordings

- *Mario Lanza: The Legendary Tenor*, (1987)
- *Christmas With Mario Lanza* (1987)
- *The Great Caruso And Other Caruso Favorites* (1989)
- *Mario Lanza Sings Songs from The Student Prince and The Desert Song* (1989)
- *The Mario Lanza Collection*, (1991)
- *Mario Lanza Live From London* (1994)
- *Mario! Lanza At His Best* (1995)
- *Mario Lanza: Opera Arias and Duets*, (1999)
- *Mario Lanza Live at Hollywood Bowl: Historical Recordings (1947 & 1951)* (2000)
- *Serenade/A Cavalcade of Show Tunes* (2004)
- *The Essential Mario Lanza* (2007)

References

[1] Vogel, Michelle. *Children of Hollywood*, McFarland (2005), p. 65
[2] Mario Lanza (I) – Biography (http://www.imdb.com/name/nm0487292/bio)
[3] Fischer, Lucy; Landy, Marcia. *Stars: The Film Reader*, Routledge (2004) p. 216
[4] Hopper, Hedda. *The Whole Truth and Nothing But*, Pyramid Books (1963) ch. 18
[5] Mannering, Derek. *Mario Lanza: Singing to the Gods*, Univ. Press of Mississippi (2005) p. xv–xvii
[6] Kimmel, Eleonora. *Altered and Unfinished Lives*, A.F.A. (2006) p. 191
[7] Mario Lanza – Photoplay, Sept. 1951 (http://www.seektress.com/photop2.htm)
[8] Great Tenors (http://jedi.cs.mcla.edu/RickCampedelli/Web Development/final/lanza.html)
[9] "Mario Lanza in Scotland | News" (http://www.operascotland.org/news/143/Mario+Lanza+in+Scotland+). Opera Scotland. . Retrieved 2012-04-14.
[10] Cesari, Armando. *Mario Lanza: An American Tragedy* (2004) Passage link (http://books.google.co.uk/books?id=2vlpWdsgQ3UC&pg=PA284&lpg=PA284&dq="Mario+Lanza"+funeral+"Frank+Sinatra"&source=web&ots=L8IRM9Y5kT&sig=PfJXo65bdE_xld7sAiNHOfRA3w4&hl=en&sa=X&oi=book_result&resnum=3&ct=result#PPA284,M1)
[11] Cesari, Armando. *Mario Lanza: An American Tragedy*, Baskerville Publishers (2004) p. 4
[12] Interview with José Carréras for New Zealand Television, 1994 (http://www.solopassion.com/node/2843)
[13] Plácido Domingo Interview with CBS, January 2009 (http://cbs3.com/specialreports/Eye.On.The.2.922970.html)
[14] (http://www.wqxr.org/#!/articles/wqxr-features/2012/jan/19/cafe-concert-joseph-calleja/)
[15] McCants, Clyde T. *American Opera Singers and Their Recordings*, McFarland (2004), p. 132
[16] "Richard Vetere Collection" (http://www.stonybrook.edu/libspecial/collections/manuscripts/vetere.shtml). Stony Brook University Special Collections & University Archives. .
[17] Palm Springs Walk of Stars by date dedicated (http://www.palmspringswalkofstars.com/web-storage/Stars/Stars dedicated by date.pdf)

Further reading

- Callinicos, Constantine. *The Mario Lanza Story* (New York, NY, 1960). (Library of Congress Catalog Card Number 60-12480)
- Lanza, Damon & Dolfi, Bob. *Be My Love: A Celebration of Mario Lanza* (Chicago, IL, 1999) (ISBN 1-56625-129-X)
- Bessette, Roland L. *Mario Lanza: Tenor In Exile* (Portland, OR) (ISBN 1-57467-044-1)

External links

- The Lanza Legend (http://www.lanzalegend.com/)
- Mario Lanza (http://www.imdb.com/name/nm487292/) at the Internet Movie Database
- Mario Lanza (http://www.ibdb.com/person.asp?ID={{{id}}}) at the Internet Broadway Database
- Mario Lanza (http://www.discogs.com/artist/Mario+Lanza) discography at Discogs
- MarioLanza.net (http://www.mariolanza.net) Hosted by Jeff Rense.
- Mario Lanza, Tenor (http://www.mariolanzatenor.com) – Essays, videos, rare recordings, discography and forum.
- History of the Tenor – Sound Clips and Narration (http://historyofthetenor.com/page.php?87)
- (http://mariolanzarevisited.blogspot.com)

John_Forsythe

John Forsythe	
 in a trailer for *The Trouble with Harry* (1955)	
Born	Jacob Lincoln FreundJanuary 29, 1918Penns Grove, New Jersey, U.S.
Died	April 1, 2010 (aged 92)Santa Ynez, California, U.S.
Cause of death	Pneumonia
Resting place	Oak Hill Cemetery, Ballard, California
Nationality	American
Education	Abraham Lincoln High School
Alma mater	University of North Carolina at Chapel Hill
Occupation	Actor
Years active	1943 – 2006
Spouse(s)	Parker McCormick (m.1939-1943; divorced) Julie Warren (m.1943-1994; her death) Nicole Carter (m.2002–2010; his death)
Children	Page Warren Forsythe Brooke Warren Forsythe Dall McCormick Forsythe
Awards	**Golden Globe Award** (1984)

John Forsythe (born **Jacob Lincoln Freund**;[1] January 29, 1918 – April 1, 2010) was an American stage, television, and film actor.[2] Forsythe starred in three television series, spanning four decades and three genres: as single playboy father, Bentley Gregg, in the sitcom *Bachelor Father* (1957–1962);[2] as the unseen millionaire, Charles Townsend,[2] on the crime drama *Charlie's Angels* (1976–1981), and as patriarch, Blake Carrington, on the soap opera *Dynasty* (1981–1989).[2] He hosted *World of Survival* (1971–1977).[2]

Early life

The eldest of three children, Forsythe was born as **Jacob Lincoln Freund** in Penns Grove, New Jersey, to Blanche Materson (née Blohm) and Samuel Jeremiah Freund, a stockbroker.[3] He was raised in Brooklyn, New York, where his father worked as a Wall Street businessman during the Great Depression of the 1930s.

He graduated from Abraham Lincoln High School in Brooklyn at the age of 16, and began attending the University of North Carolina at Chapel Hill.[4] In 1936 at the age of eighteen, he took a job as the public address announcer for Brooklyn Dodgers games at Ebbets Field, confirming a childhood love of baseball.[5]

Movie career and Army service

Despite showing initial reluctance, Forsythe began an acting career at the suggestion of his father. He met actress, Parker McCormick (December 1918 – July 1980), and the couple married in 1939; they had a son, Dall (born in 1943), and divorced in 1943. As a bit player for Warner Brothers, Forsythe successfully appeared in several small parts. As a result he was given a small role in *Destination Tokyo* (1943). Leaving his movie career for service in World War II, he appeared in the U.S. Army Air Forces play and film *Winged Victory*, then worked with injured soldiers who had developed speech problems.

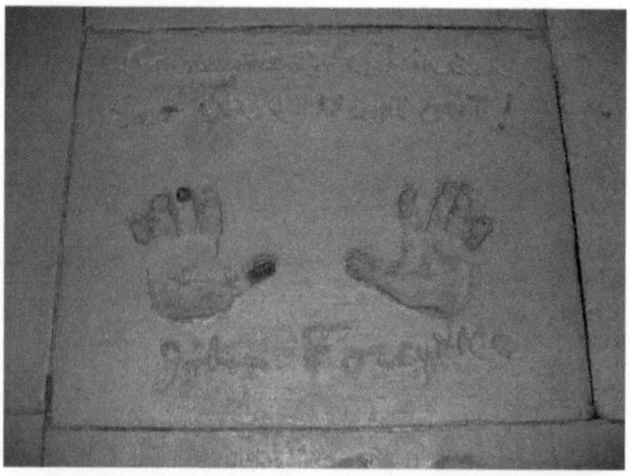
The handprints of John Forsythe in front of The Great Movie Ride at Walt Disney World's Disney's Hollywood Studios theme park.

Also in 1943, Forsythe met Julie Warren, initially a theatre companion, but later a successful actress in her own right, landing a role on Broadway in *Around the World*. Warren became Forsythe's second wife and in the early 1950s the marriage produced two daughters – Page[6] and Brooke.[7] In 1947, Forsythe joined the initial class of the soon-to-be prestigious Actors Studio, where he met other promising young actors including Marlon Brando and Julie Harris. During this time he appeared on Broadway in *Mister Roberts* and *The Teahouse of the August Moon*.

In 1955 Alfred Hitchcock cast Forsythe in the movie *The Trouble with Harry*, with Shirley MacLaine in her first movie appearance, for which she won a Golden Globe.

Television work

Throughout the 1950s, Forsythe successfully appeared in the new medium and regularly on all the networks, especially as a guest star. For example, during this period, Forsythe notably appeared on the popular anthology *Alfred Hitchcock Presents* in an the episode entitled "Premonition" opposite Cloris Leachman.

Star on the Hollywood Walk of Fame at 6549 Hollywood Blvd.

Bachelor Father

In 1957, he took a leading role in the situation comedy *Bachelor Father* for CBS as Bentley Gregg, a playboy lawyer who has to become a father to his niece Kelly (played by Noreen Corcoran), upon the death of her biological parents. The show was an immediate ratings hit and moved to NBC the following season and to ABC in the fall of 1961. On various episodes Forsythe worked with such up-and-coming actresses as Mary Tyler Moore, Barbara Eden, Donna Douglas, Sally Kellerman, Sue Ane Langdon, and a teenaged Linda Evans. During the 1961 season, *Bachelor Father* moved to ABC, but was cancelled that season due to declining ratings.

After *Bachelor Father*

In the early 1960s, Forsythe returned to acting in movies including *Kitten with a Whip* (1964), *Madame X* (1966) and *In Cold Blood* (1967). In 1964 he starred in the made for TV Movie *See How They Run* which is notable for being the first movie made for television. He also attempted two new television programs: *The John Forsythe Show* on NBC with Guy Marks, Elsa Lanchester, Ann B. Davis, Peggy Lipton, and Forsythe's two young daughters, Page and Brooke (1965–1966), and *To Rome With Love* on CBS (1969–1971) with co-star Walter Brennan. Between 1971 and 1977, Forsythe served as narrator on the syndicated nature series, *The World of Survival*. He was also the announcer for Michelob beer commercials from the 70s through about 1985, notably during the "Weekends were made for Michelob" era.

Charlie's Angels

Forsythe began a 13-year association with Aaron Spelling in 1976, cast in the role of mysterious unseen millionaire private investigator Charles Townsend in the crime drama *Charlie's Angels* (1976–1981). The show starred Kate Jackson, Jaclyn Smith, and Farrah Fawcett, making stars of all three but catapulting Fawcett to iconic status. Forsythe introduces the series' concept during its opening credits:

> Once upon a time, three little girls went to the police academy, where they were each assigned very hazardous duties. But I took them away from all that, and now they work for me. My name is Charlie.

On every episode, Forsythe, as Townsend, is heard over a speaker phone, instructing the eponymous Angels on their mission for the episode. Character actor David Doyle played Charlie's on-screen liaison to the Angels, who in later seasons included Cheryl Ladd, Shelley Hack, and Tanya Roberts. *Charlie's Angels* was a huge success, and was shown in more than 90 countries.

Forsythe became the highest paid actor on television on a per-hour basis: while the show's on-camera stars often worked 15-hour days five days a week, with a couple hours just for hair and makeup, Forsythe's lines for an entire episode would be recorded in a sound studio in a matter of minutes, after which he would have lunch in the network's commissary and then head for the track. During this period, Forsythe invested a lot of money in thoroughbred racing, a personal hobby. Gaining respect with the celebrity thoroughbred circuit, he served on the Board of Directors at the Hollywood Park Racetrack since 1972, and was on the committee for more than 25 years.

Following heart problems, Forsythe underwent quadruple coronary artery bypass surgery in 1979. This was so successful that he not only returned to work on *Charlie's Angels*, he also appeared in the two-time Academy Award-nominated motion picture *...And Justice for All* later that year as Judge Henry T. Fleming, the film's main antagonist, an evil corrupt judge who despises Al Pacino's lawyer character.

Dynasty

In 1981, nearing the end of *Charlie's Angels*, Forsythe was selected as a last minute replacement for George Peppard in the role of conniving patriarch Blake Carrington in *Dynasty*. Another Aaron Spelling production, *Dynasty* was ABC's answer to the highly successful CBS series *Dallas*. Between 1985 and 1987, Forsythe also appeared as Blake Carrington in the short-lived spin-off series *The Colbys*.

The series reunited Forsythe with one-time *Bachelor Father* guest star Linda Evans, who would play Blake's wife, Krystle. During the run of the series, Forsythe, Evans and another co-star Joan Collins, who played Blake's ex-wife Alexis, promoted the *Dynasty* line of fragrances.[8] *Dynasty* came to an end in 1989, after a total of nine seasons. Forsythe was the only actor to appear in all 220 episodes.

Forsythe was nominated for Emmy Awards three times between 1982 and 1984 for "Outstanding Lead Actor in a Drama Series" but did not win. He was also nominated six times for Golden Globe Awards, winning twice. He was nominated five times for the *Soap Opera Digest* Awards, also winning twice.

The Powers That Be

In 1992, after a three-year absence, Forsythe returned to series television starring in Norman Lear's situation comedy, *The Powers That Be* for NBC, co-starring Holland Taylor, Peter MacNicol, Valerie Mahaffey and David Hyde-Pierce. He was reunited with one-time co-star Cloris Leachman, who, although eight years his junior, played his mother in one episode. The show was canceled after only one year.

Post-1990s work and life

Forsythe's wife of 51 years, Julie Warren (October 20, 1919 — August 15, 1994), died at the age of 74,[9] in her hospital room, after Forsythe made the decision to turn off her life-support system. She had been in a coma following severe breathing difficulties. In July 2002, Forsythe married businesswoman Nicole Carter (May 27, 1941 - May 11, 2010)[10] at Ballard Country Church;[11] they remained married until his death. Forsythe is survived by his son, two daughters, six grandchildren, and four great-grandchildren.

Forsythe reprised his role as Charlie for the film version of *Charlie's Angels* (2000) and its sequel *Charlie's Angels: Full Throttle* (2003), but then retired from acting. Besides spending time with his family, he enjoyed ownership of an art gallery. In 2005 actor Bartholomew John portrayed Forsythe in *Dynasty: The Making of a Guilty Pleasure*, a fictionalized television movie based on the creation and behind the scenes production of *Dynasty*.[12]

On May 2, 2006, Forsythe appeared with *Dynasty* co-stars Linda Evans, Joan Collins, Pamela Sue Martin, Al Corley, Gordon Thomson and Catherine Oxenberg in *Dynasty Reunion: Catfights & Caviar*. The one-hour reunion special of the former ABC series aired on CBS. Forsythe appeared each year to read children's fiction during the annual Christmas program near his home at the rural resort community of Solvang, California.

Forsythe was treated for colorectal cancer in the fall of 2006.[13] Surgery was reportedly successful and his cancer was considered to be in remission.[14]

Death

Forsythe died on April 1, 2010, from pneumonia in Santa Ynez, California, at the age of 92.[2] [15] [16] His widow, Nicole, died 40 days later.[11] He was interred at Oak Hill Cemetery in Santa Barbara, California.[17]

Thoroughbred racing

Forsythe owned and bred Thoroughbred racehorses for many years and was a member of the Board of Directors of Hollywood Park Racetrack. Among his successes, in partnership with film producer Martin Ritt he won the 1976 Longacres Mile with Yu Wipi. With partner Ken Opstein, he won the 1982 Sixty Sails Handicap with Targa, and the 1993 La Brea Stakes with a daughter of Targa, Mamselle Bebette, which he raced under the name of his Big Train Farm, a stable he named for Hall of Fame baseball pitcher, Walter Johnson,[18] In the 1980s, John Forsythe served as the regular host for the annual Eclipse Awards. He was the recipient of the 1988 Eclipse Award of Merit for his contribution in promoting the sport of Thoroughbred racing.[19]

Filmography

- *Northern Pursuit* (1943)
- *Destination Tokyo* (1943)
- *The Captive City* (1952)
- *It Happens Every Thursday* (1953)
- *The Glass Web* (1953)
- *Escape from Fort Bravo* (1953)
- *American Harvest (Revised Edition)* (1955) (short subject) (narrator)
- *The Trouble with Harry* (1955)
- *The Ambassador's Daughter* (1956)
- *Everything But the Truth* (1956)
- *Dubrowsky* (1959)
- *Kitten with a Whip* (1964)
- *Madame X* (1966)
- *In Cold Blood* (1967)
- *Silent Treatment* (1968)
- *Murder Once Removed* (1971)
- *Topaz* (1969)
- *The Happy Ending* (1969)
- *Goodbye and Amen* (1977)
- *...And Justice for All* (1979)
- *Scrooged* (1988)
- *Stan and George's New Life* (1991)
- *We Wish You a Merry Christmas* (1999) (voice) (direct-to-video)
- *Charlie's Angels* (2000)
- *Charlie's Angels: Full Throttle* (2003)

Television work

- *Alfred Hitchcock Presents* (1955)
- *Bachelor Father* (1957–1962)
- *The Ford Show, Starring Tennessee Ernie Ford* (June 9, 1960)
- *The Alfred Hitchcock Hour* (1962)
- *See How They Run* (1964)
- *The John Forsythe Show* (1965–1966)
- *A Bell for Adano* (1967)
- *Shadow on the Land* (1968)
- *To Rome With Love* (1969–1971)
- *Murder Once Removed* (1971)
- *The World of Survival* (1971–1977) (narrator)
- *The Letters* (1973) (unsold pilot)
- *Lisa, Bright and Dark* (1973)
- *Cry Panic* (1974)
- *The Healers* (1974)
- *Terror on the 40th Floor* (1974)
- *The Deadly Tower* (1975)
- *Charlie's Angels* (1976–1981)
- *Amelia Earhart* (1976)
- *Tail Gunner Joe* (1977)
- *Emily, Emily* (1977)
- *Never Con a Killer* (1977) (pilot for *The Feather and Father Gang*)
- *Cruise Into Terror* (1978)
- *With This Ring* (1978)
- *The Users* (1978)
- *A Time for Miracles* (1980)
- *Dynasty* (1981–1989)
- *Sizzle* (1981)
- *Mysterious Two* (1982)
- *The Love Boat* (1983)
- *The Colbys* (1985–1986)
- *On Fire* (1987)
- *Miss Universe Pageant* (1989)
- *Opposites Attract* (1990)
- *Dynasty: The Reunion* (1991)
- *The Powers That Be* (1992–1993)
- *I Witness Video* (host from 1993–1994)
- *People's Century* (1995) (miniseries) (narrator in U.S. version)
- *Dynasty Reunion: Catfights & Caviar* (2006)

References

[1] "John Forsythe dead at 92" (http://www.rbr.com/tv-cable/23032.html?print?print). RBR.comTVBR.com. 2010-04-02. . Retrieved 2010-09-22.

[2] Adam Bernstein (April 3, 2010). "John Forsythe dead; starred in 'Dynasty,' 'Bachelor Father'" (http://www.washingtonpost.com/wp-dyn/content/article/2010/04/02/AR2010040202419.html). *The Washington Post*. . Retrieved April 3, 2010.

[3] "John Forsythe Biography" (http://www.filmreference.com/film/43/John-Forsythe.html) Film Reference

[4] Staff. "Biography for John Forsythe" (http://www.tcm.com/tcmdb/participant.jsp?spid=539265), *Turner Classic Movies*. Retrieved September 23, 2009. "Attending Brooklyn's Abraham Lincoln High School, he came of age, like countless Brooklyn youngsters, a fan of the Brooklyn Dodgers and devoted his extracurricular activities to sports."

[5] A Tribute to John Forsythe website (http://www.johnforsythe.info/biography.php)

[6] Page Forsythe (http://www.imdb.com/name/nm0287110/) at the Internet Movie Database

[7] Brooke Forsythe (http://www.imdb.com/name/nm0287087/) at the Internet Movie Database

[8] "TVAcres.com" (http://www.tvacres.com/perfumes_forever.htm). .

[9] Julie Forsythe's lifespan per Social Security Death Index (http://ssdi.rootsweb.ancestry.com)

[10] http://www.legacy.com/obituaries/latimes/obituary.aspx?n=nicole-carter-forsythe&pid=142979575

[11] legacy.com obituaries (http://www.legacy.com/obituaries/syvnews/obituary.aspx?n=nicole-forsythe-carter&pid=142876746)

[12] "*Dynasty: The Making of a Guilty Pleasure*: Credits" (http://www.der-denver-clan.de/de/dynasty_behind.207.html). Der-denver-clan.de. . Retrieved February 27, 2009.

[13] "Dynasty star treated for cancer" (http://news.bbc.co.uk/2/low/entertainment/6046762.stm). BBC News. October 13, 2006.

[14] http://peabodyweb.com/apps/v2_memorials_ob_detail.asp?did=1

[15] Anita Gates (April 2, 2010). "John Forsythe, 'Dynasty' Actor, Is Dead at 92" (http://www.nytimes.com/2010/04/03/arts/television/03forsythe.html). *The New York Times*. . Retrieved April 2, 2010.

[16] Claudia Luther (April 3, 2010). "John Forsythe dies at 92; actor known for roles on TV series 'Bachelor Father,' 'Charlie's Angels' and 'Dynasty'" (http://www.latimes.com/news/la-me-john-forsythe3-2010apr03,0,5912203,full.story). *Los Angeles Times*. . Retrieved April 3, 2010.

[17] http://www.findagrave.com/cgi-bin/fg.cgi?page=gr&GRid=50566155

[18] Grahame L. Jones (December 31, 2005). "This Horse Was Fair Game for the 'King'" (http://articles.latimes.com/2005/dec/31/sports/sp-briefing31). *Los Angeles Times*. . Retrieved 2010-04-03.

[19] "Eclipse memories" (http://sports.espn.go.com/sports/horse/columns/story?columnist=paulick_ray&id=3861559). *ESPN*. 2009-01-26. . Retrieved 2010-04-03.

External links

- John Forsythe (http://www.imdb.com/name/nm1234/) at the Internet Movie Database
- John Forsythe Debonair Dynasty Star Dies (http://www.trendpk.com/john-forsythe-debonair-dynasty-star-dies/30290.html)
- Official site (http://www.johnlforsythe.com/)
- Tributes (http://www.lastingtribute.co.uk/tribute/forsythe/3289652/Memories)
- John Forsythe Archive of American Television interview (2000) (http://emmytvlegends.org/interviews/people/john-forsythe)
- John Forsythe (http://www.findagrave.com/cgi-bin/fg.cgi?page=gr&GRid=50566155) at *Find a Grave*

Karl_Malden

Karl Malden	
 Malden in 2009	
Born	Mladen George Sekulovich *Младен Ђорђе Секуловић*March 22, 1912Chicago, Illinois, U.S.
Died	July 1, 2009 (aged 97)Brentwood, Los Angeles, California, U.S.
Resting place	Westwood Village Memorial Park Cemetery
Residence	Brentwood, California
Nationality	American
Education	Emerson School for Visual and Performing Arts
Alma mater	DePaul University
Occupation	Actor
Years active	1936–2000
Home town	Gary, Indiana (raised)
Spouse(s)	Mona Greenberg (1938–2009)
Children	Mila Malden, Carla Malden
Parents	Petar Sekulovic, Minnie (née Sebera) Sekulovich
Awards	*See Awards and recognition*

Karl Malden (born **Mladen Djordje Sekulovich**; Serbian Cyrillic Младен Ђорђе Секуловић; March 22, 1912 – July 1, 2009)[1] [2] [3] was a Serbian American[4] actor. In a career that spanned more than seven decades, he performed in such classic films as *A Streetcar Named Desire*, for which he won the Academy Award for Best Supporting Actor; *On the Waterfront*; and *One-Eyed Jacks*. Among other notable film roles were Archie Lee Meighan in *Baby Doll*, Zebulon Prescott in *How the West Was Won*, and General Omar Bradley in *Patton*. His best-known role, though, was on television as Lt. Mike Stone on the 1970s crime drama, *The Streets of San Francisco*. During the 1970s and 1980s, he was also the spokesman for American Express.

Early life

Malden, the eldest of three brothers, was born in Chicago, Illinois and raised in Gary, Indiana.[1] His Serbian father, Petar Sekulović (1886–1975), worked in the steel mills and as a milkman, and his mother, Minnie (née Sebera) Sekulovich (1892–1995), was a Czech seamstress and actress.[5] The Sekulovich family roots trace back to Podosoje near the city of Bileća in Bosnia and Herzegovina, more specifically, in the eastern portion of Herzegovina now resting within Republika Srpska. Malden spoke only the Serbian language until he was in kindergarten and was fluent in the language until his death. Malden's father had a passion for music, and organized a choir. As a teenager, Malden joined the Karageorge Choir. In addition, his father produced Serbian plays at his church and taught acting. A young Malden took part in many of these plays, which included a version of *Jack and the Beanstalk*, but mostly centered on the community's Serbian heritage. In high school, he was a popular student and the star of the basketball team (according to his autobiography, Malden broke his nose twice while playing, taking elbows to the face and resulting in his trademark bulbous nose).[6] He participated in the drama department, and was narrowly elected senior class president. Among other roles, he played Pooh Bah in *The Mikado*. After graduating from Emerson High School in 1931 with high marks, he briefly planned to leave Gary for Arkansas, where he hoped to win an athletic scholarship, but college officials did not admit him owing to his refusal to play any sport besides basketball. From 1931 until 1934, he worked in the steel mills, as had his father.

He changed his name from **Mladen Sekulovich** to **Karl Malden** at age 22. He anglicized his first name by swapping its letters "l" and "a" and making the result, Malden, his last name; then he proceeded to take his grandfather's first name of Karl and make it his first name.[7] This was because the first theatre company he was in wanted him to shorten his name for its marquee. He thought they wanted to fire him and were using his name as an excuse; although that wasn't the case, he still changed his name so as to give them no excuse.

Malden often found ways to say "Sekulovich" in films and television shows in which he appears. For example, as General Omar Bradley in *Patton*, as his troops slog their way through enemy fire in Sicily, Malden says "Hand me that helmet, Sekulovich" to another soldier. In *Dead Ringer*, as a police detective in the squad room, Malden tells another detective: "Sekulovich, gimme my hat." In *Fear Strikes Out*, Malden, playing Jimmy Piersall's father John, introduces Jimmy to a baseball scout named Sekulovich. In *Birdman of Alcatraz*, as a prison warden touring the cell block, Malden recites a list of inmates' names, including Sekulovich. (Malden's father was not pleased, as he told his son "Mladen, no Sekulovich has ever been in prison!") In *On the Waterfront*, in which Malden plays the priest, among the names of the officers of Local 374 called out in the courtroom scene is Mladen Sekulovich, Delegate. Perhaps the most notable usage of his real name, however, was in the TV series *The Streets of San Francisco*. Malden's character in the program, Mike Stone, employed a legman (played by Art Metrano) with that name, who did various errands.

Education and early stage work

In September 1934, Malden decided to leave his home in Gary, Indiana, to pursue formal dramatic training at the Goodman School (later part of DePaul University), then associated with the Goodman Theater in Chicago. Although he had worked in the steel mills in Gary for three years, he had helped support his family, and was thus unable to save enough money to pay for his schooling. Making a deal with the director of the program, he gave the institute the little money that he did have, with the director agreeing that, if Malden did well, he would be rewarded with a full scholarship. He won the scholarship. When Malden performed in the Goodman's children's theater, he wooed the actress Mona Greenberg (stage name: Mona Graham), who married him in 1938. He graduated from the Chicago Art Institute in 1937. Soon after, without work and without money, Malden returned to his hometown.

Acting career circa World War II

He eventually traveled to New York City, and first appeared as an actor on Broadway in 1937. He did some radio work and in a small role made his film debut in *They Knew What They Wanted.* He also joined the Group Theatre, where he began acting in many plays and was introduced to a young Elia Kazan, who would later work with him on *A Streetcar Named Desire* (1951) and *On the Waterfront* (1954).

His acting career was interrupted by World War II, during which he served as a noncommissioned officer in the 8th Air Force. While in the service, he was given a small role in the United States Army Air Forces play and film *Winged Victory*. After the war ended in 1945, he resumed his acting career, playing yet another small supporting role in the Maxwell Anderson play *Truckline Cafe*, with a then-unknown Marlon Brando. He was given a co-starring role in the Arthur Miller play *All My Sons* with the help of director Elia Kazan. With that success, he then crossed over into steady film work.

Film career: 1950s to 1970s

Malden resumed his film acting career in the 1950s, starting with *The Gunfighter* (1950) and *Halls of Montezuma* (1950). The following year, he was in *A Streetcar Named Desire* (1951), playing Mitch, Stanley Kowalski's best friend who starts a romance with Blanche DuBois (Vivien Leigh). For this role, he won an Academy Award for Best Supporting Actor.

Other films during this period include Alfred Hitchcock's *I Confess* with Montgomery Clift and Anne Baxter (1953), *On the Waterfront* (1954), where he played a priest who influenced Terry Malloy (Marlon Brando) to testify against mobster-union boss Johnny Friendly (Lee J. Cobb). In *Baby Doll* (1956), he played a power-hungry sexual man who had been frustrated by a teenage wife. He starred in dozens of films from the late 1950s to the early 1970s, such as *Fear Strikes Out* (1957), *Pollyanna* (1960), *Birdman of Alcatraz* (1962), *Gypsy* (1962), *How the West Was Won* (1962), *The Cincinnati Kid* (1965), and *Patton* (1970), playing General Omar Bradley. After *Summertime Killer* (1972), he appeared in the made-for-television film *The Hijacking of the Achille Lauro* (1989) (as Leon Klinghoffer).

Malden's wife, Mona, the former Mildred Greenberg, graduated from Roosevelt High School in Emporia, Kansas where she attended Kansas State Teachers College, now Emporia State University. He first visited the campus with her in 1959 and was impressed by the ESU Summer Theatre. He returned in the summer of 1964 to teach, working with the actors in the company. Upon leaving, he gave his honorarium to establish the Karl Malden Theater Scholarship still given today.[8]

In 1963, he was a member of the jury at the 13th Berlin International Film Festival.[9]

Television work

In 1972, Malden was approached by producer Quinn Martin about starring as Lt. Mike Stone in *The Streets of San Francisco.* Although the concept originated as a made-for-television movie, ABC quickly signed on to carry it as a series. Martin hired Michael Douglas to play Lt. Stone's young partner, Inspector Steve Keller.

Malden's father was delighted about this series being in San Francisco, as he had intended to settle in that city, but had to change his plans as he'd arrived on the day of the 1906 San Francisco earthquake.

On *Streets*, Malden played a widowed veteran cop with more than 20 years of experience who is paired with a young officer recently graduated from college. During its first season, it was a ratings winner among many other 1970s crime dramas, and served as ABC's answer to such shows as *Hawaii Five-O*, *Adam-12*, *Ironside*, *Barnaby Jones*, *Kojak*, *McMillan & Wife*, *Police Woman*, *The Rockford Files*, and *Switch.*

During the second season, production shifted from Los Angeles to San Francisco. For his work as Lt. Stone, Malden was nominated for a Primetime Emmy Award for Outstanding Lead Actor - Drama Series four times between 1974 and 1977, but never won. After two episodes in the fifth season, Douglas left the show to act in movies; Douglas had also produced the film *One Flew Over the Cuckoo's Nest* in 1975. Lt. Stone's new partner was Inspector Dan Robbins, played by Richard Hatch. The show took a ratings nosedive, and ABC canceled it after five seasons and 119 episodes.

In 1980, Malden starred in *Skag*, an hour-long drama that focused on the life of a foreman at a Pittsburgh steel mill. Malden described his character, Pete Skagska, as a simple man trying to keep his family together. The pilot episode for the series had Skag temporarily disabled by a stroke, and explored the effects it had on his family and co-workers. While *Skag* met with poor ratings, critics praised it, in instances there were even full page ads taken out in newspapers in an attempt to keep the program from being taken off the air. Nevertheless the series was canceled after several episodes.

In 1987, he was the host/narrator for the second and third television specials that later became the long-running series *Unsolved Mysteries*.

Malden's last role in film or television was in 2000 in the highly acclaimed first season episode of *The West Wing* titled "Take This Sabbath Day". Malden portrayed a Catholic priest and used the same Bible he had used in *On the Waterfront*.

Other work

Malden famously delivered the line "Don't leave home without them!" in a series of U.S. television commercials for American Express Travelers Cheques in the 1970s and 1980s. He also advertised the American Express Card, with the famous opening line, "Do you know me?"

Malden was a member of the United States Postal Service's 16-member Citizens' Stamp Advisory Committee, which meets to review recommendations for U.S. commemorative postage stamps.[10] [11]

Personal life

On December 18, 1938, Malden married **Mona Greenberg**, who survives him. Their marriage was one of the longest in Hollywood's history,[12] lasting more than 70 years. In addition to his wife, Malden is survived by daughters Mila and Carla; his sons-in-law; three granddaughters, Alison, Emily, and Cami; plus four great-grandchildren, Mila, Stella, Charlie, and Thomas Karl.[13]

In 1997, Malden published his autobiography, *When Do I Start?*, written with his daughter Carla.[6]

Death

Malden died at his home in the Brentwood section of Los Angeles on July 1, 2009 at the age of 97. He is said to have died of natural causes. Malden's manager said, "It could be many things. I mean, he was 97 years old!" He is said to have been in poor health for several years.[14]

Malden's friend and former co-star Michael Douglas wrote a tribute to Malden for *Time* magazine's "Milestones" section.

He is buried at the Westwood Village Memorial Park Cemetery, Westwood, California, United States.[15]

Karl Malden's grave

Awards and recognition

Malden won the 1951 Academy Award for Best Supporting Actor for *A Streetcar Named Desire* and was nominated in 1954 for his supporting role in *On the Waterfront*. Malden was a past president of the Academy of Motion Picture Arts and Sciences. In October 2003, he was named the 40th recipient of the Screen Actors' Guild's Life Achievement Award for career achievement and humanitarian accomplishment.

In 1985, he was awarded an Emmy Award for Outstanding Supporting Actor in a Limited Series for his performance as Freddy Kassab in *Fatal Vision*. During the same year, he was also awarded an honorary doctor degree in fine arts by Emporia State University.[8]

In May 2001, Malden received an honorary degree, Doctor of Humane Letters, from Valparaiso University. Michael Douglas, his former "The Streets of San Francisco" co-star, presented Malden with a Lifetime Achievement Award from the Screen Actors Guild on February 22, 2004.[16] On November 11, 2004, Douglas also presented Malden with the Monte Cristo Award of the Eugene O'Neill Theater Center in Waterford, Connecticut, which is given for "distinguished careers exemplifying Eugene O'Neill's standard of excellence and pioneering spirit." Among other past winners were Jason Robards, Zoe Caldwell, Edward Albee, August Wilson and Brian Dennehy.

On November 12, 2005, the United States House of Representatives authorized the United States Postal Service to rename the Los Angeles Barrington Postal Station as the Karl Malden Postal Station in honor of Malden's achievements. The bill, H.R. 3667, was sponsored by Representatives Henry Waxman and Diane Watson.

For his contribution to the film industry, Malden has a star on the Hollywood Walk of Fame at 6231 Hollywood Blvd. In 2005, he was inducted into the Western Performers Hall of Fame at the National Cowboy & Western Heritage Museum in Oklahoma City, Oklahoma.[17]

Filmography

Karl Malden in the trailer for *I Confess* (1953)

Karl Malden with Marlon Brando in the trailer for *On the Waterfront* (1954)

- *They Knew What They Wanted* (1940)
- *Winged Victory* (1944)
- *13 Rue Madeleine* (1947)
- *Boomerang* (1947)
- *Kiss of Death* (1947)
- *The Gunfighter* (1950)
- *Where the Sidewalk Ends* (1950)
- *Halls of Montezuma* (1951)
- *A Streetcar Named Desire* (1951)
- *The Sellout* (1952)
- *Diplomatic Courier* (1952)
- *Operation Secret* (1952)
- *Ruby Gentry* (1952)
- *I Confess* (1953)
- *Take the High Ground!* (1953)
- *Phantom of the Rue Morgue* (1954)
- *On the Waterfront* (1954)
- *Baby Doll* (1956)
- *Fear Strikes Out* (1957)
- *Bombers B-52* (1957)
- *Time Limit* (1957) (director)
- *The Hanging Tree* (1959)
- *Pollyanna* (1960)
- *The Great Impostor* (1961)
- *One-Eyed Jacks* (1961)
- *Parrish* (1961)
- *All Fall Down* (1962)
- *Birdman of Alcatraz* (1962)
- *How the West Was Won* (1962)
- *Gypsy* (1962)
- *Come Fly with Me* (1963)
- *Dead Ringer* (1964)
- *Cheyenne Autumn* (1964)
- *The Cincinnati Kid* (1965)
- *Nevada Smith* (1966)
- *Murderers' Row* (1966)
- *Hotel* (1967)
- *The Adventures of Bullwhip Griffin* (1967)
- *Billion Dollar Brain* (1967)
- *Blue* (1968)
- *Hot Millions* (1968)
- *Patton* (1970)
- *The Cat o' Nine Tails* (1971)
- *Wild Rovers* (1971)

- *The Streets of San Francisco* (1972 - 1977)
- *Summertime Killer* (1972)
- *Beyond the Poseidon Adventure* (1979)
- *Meteor* (1979)
- *Miracle on Ice* (1981)
- *Twilight Time* (1982)
- *The Sting II* (1983)
- *Dario Argento's World of Horror* (1985) (documentary)
- *Billy Galvin* (1986)
- *Nuts* (1987)
- *My Father, My Son* (1988)
- *The Hijacking of the Achille Lauro* (1989)
- *Absolute Strangers* (1991)
- *Back to the Streets of San Francisco* (1992)
- *Vanished Without a Trace* (TV Movie 1992)
- *Broadway: The Golden Age, by the Legends Who Were There* (2003)

Karl Malden with Eva Marie Saint in the trailer for *On the Waterfront* (1954)

Karl Malden as *Father Barry* in the trailer for *On the Waterfront* (1954)

Television miniseries

- *Fatal Vision* (1984)
- *Alice in Wonderland* (1985)

Notes

[1] McLellan, Dennis (2009-07-01). "Oscar-winning actor Karl Malden dies at 97" (http://www.latimes.com/news/obituaries/la-me-karl-malden2-2009jul02,0,5658128.story). Los Angeles Times. . Retrieved 2009-07-01.

[2] Ebright, Olsen (2009-07-01). "Legendary Actor Karl Malden Dead at 97" (http://www.nbclosangeles.com/news/entertainment/Actor-Karl-Malden-Dead-at-97.html). *NBC Los Angeles* (online). .

[3] Potempa, Philip (2007-12-02). "Hard work and dreams: Karl Malden remembers his roots" (http://www.nwitimes.com/sports/article_8c1ea98f-96fd-5bbf-b93a-2d1ec622cef0.html). *The Times of Northwest Indiana* (online). .

[4] Karl Malden - SAG Life Achievement Award in Los Angeles, 2004 (http://www.youtube.com/watch?v=eBS0vC90kxM). Retrieved 5 April 2012.

[5] "Karl Malden Biography (1914?-)" (http://www.filmreference.com/film/4/Karl-Malden.html). Filmreference.com. . Retrieved 2012-03-09.

[6] *When Do I Start?: A Memoir.* (Karl Malden, with Carla Malden). Riverside, New Jersey, U.S.: Simon & Schuster, 1997. ISBN 978-0-6848-4309-4.

[7] "Oscar-winning actor Karl Malden dead at 97 - CNN.com" (http://www.cnn.com/2009/SHOWBIZ/TV/07/01/karl.malden.obit/index.html). *CNN.* July 1, 2009. . Retrieved May 7, 2010.

[8] "Actor Karl Malden was supporter of theater at ESU" (http://www.emporiagazette.com/news/2009/jul/02/actor_karl_malden_was_supporter_theater_esu). Emporia Gazette. 2009-07-02. . Retrieved 2009-07-02.

[9] "Berlinale: Juries" (http://www.berlinale.de/en/archiv/jahresarchive/1963/04_jury_1963/04_Jury_1963.html). *berlinale.de*. . Retrieved 2010-02-13.

[10] *USPS: Citizens' Stamp Advisory Committee* (http://www.usps.com/communications/organization/csac.htm). USPS.com.

[11] See this for a picture of the Yip Harburg commemorative Malden was partly responsible for and a description of the campaign to have it issued. (http://thefastertimes.com/theatertalk/2009/11/01/bustin-with-bliss-5q4-ernie-harburg/)

[12] "Oscar-winning actor Karl Malden dies aged 97" (http://uk.news.yahoo.com/4/20090701/ten-oscar-winning-actor-karl-malden-dies-ea4616c.html). Yahoo!. .

[13] Stein, Ruthe (July 2, 2009). "KARL MALDEN 1912-2009 / 'Streetcar' to 'Streets' - actor who could do it all" (http://www.sfgate.com/cgi-bin/article.cgi?f=/c/a/2009/07/01/MN8O11IQEF.DTL). *The San Francisco Chronicle*. .

[14] "Hollywood actor Karl Malden dies" (http://news.bbc.co.uk/2/hi/entertainment/8129710.stm). *BBC News*. July 2, 2009. . Retrieved May 7, 2010.

[15] Karl Malden (http://www.findagrave.com/cgi-bin/fg.cgi?page=gr&GRid=38952159) at *Find a Grave*

[16] Karl Malden receives a lifetime achievement award (Screen Actors Guild). (http://www.youtube.com/watch?v=eBS0vC90kxM) Retrieved 21 Jan 2011.

[17] "Great Western Performers" (http://www.nationalcowboymuseum.org/info/awards-hof/Western-Performers.aspx). Nationalcowboymuseum.org. . Retrieved 2012-03-09.

External links

- Karl Malden (http://www.imdb.com/name/nm1500/) at the Internet Movie Database
- Karl Malden (http://www.ibdb.com/person.asp?ID=51153) at the Internet Broadway Database

Lee_J._Cobb

Lee J. Cobb	
 circa 1960s	
Born	Leo JacobDecember 8, 1911New York City, New York, U.S.
Died	February 11, 1976 (aged 64)Woodland Hills, California, U.S.
Occupation	Actor
Spouse(s)	Helen Beverley (m.1940) Mary Hirsch (m. 1957 – 1976)(his death)

Lee J. Cobb (December 8, 1911 – February 11, 1976) was an American actor.[1] He is best known for his performance in *12 Angry Men* (1957), his Academy Award-nominated performance in *On the Waterfront* (1954), and one of his last films, *The Exorcist* (1973). He also played the role of Willy Loman in the original Broadway performance of Arthur Miller's 1949 play *Death of a Salesman* under the direction of Elia Kazan. On television, Cobb costarred in the first four seasons of the popular, long-running western series The Virginian.

Background

Born **Leo Jacob** in New York City to a Jewish family of Russian and Romanian extraction.[2] He grew up in The Bronx, New York, on Wilkins Avenue, near Crotona Park. His parents were Benjamin (Benzion) Jacob, a compositor for a foreign-language newspaper, and Kate Neilecht.[3] Cobb studied at New York University before making his film debut in *The Vanishing Shadow* (1934). He joined the Manhattan-based Group Theatre in 1935.[4]

Career

Cobb did summer stock at Pine Brook Country Club located in the countryside of Nichols, Connecticut, in the 1930s and early 1940s. Pine Brook was the summer home of the Group Theatre (New York) from 1931 until the 1940s.[5] During World War II Cobb served in the First Motion Picture Unit of the Army Air Force.[6]

Cobb entered films in the 1930s, successfully playing middle-aged and even older men while he was still a youth. He was cast as the Kralahome in the 1946 non-musical film *Anna and the King of Siam*. He also played the sympathetic doctor in *The Song of Bernadette* and appeared as James Coburn's supervisor in the spy spoofs *In Like Flint* and *Our Man Flint*. He reprised his role of Willy Loman in the 1966 CBS television adaptation of *Death of a Salesman*, which included Gene Wilder, James Farentino, Bernie Kopell and George Segal. Cobb was nominated for an Emmy Award for the performance. Mildred Dunnock, who had co-starred in both the original stage version and the 1951 film version, again repeated her role as Linda, Willy's devoted wife.

Cobb was accused of being a possible Communist in testimony before the House Un-American Activities Committee because of his involvement in left-wing political causes and his support of political and charitable organizations alleged to be Communist fronts. He was called to testify before HUAC but refused to do so for two years until, with his career threatened by the blacklist, he relented in 1953 and gave testimony in which he named 20 people as former members of the Communist Party USA.

Later, Cobb explained why he "named names" saying:

With William Holden in *Golden Boy* (1939)

> When the facilities of the government of the United States are drawn on an individual it can be terrifying. The blacklist is just the opening gambit—being deprived of work. Your passport is confiscated. That's minor. But not being able to move without being tailed is something else. After a certain point it grows to implied as well as articulated threats, and people succumb. My wife did, and she was institutionalized. The HUAC did a deal with me. I was pretty much worn down. I had no money. I couldn't borrow. I had the expenses of taking care of the children. Why am I subjecting my loved ones to this? If it's worth dying for, and I am just as idealistic as the next fellow. But I decided it wasn't worth dying for, and if this gesture was the way of getting out of the penitentiary I'd do it. I had to be employable again.
>
> — Interview with Victor Navasky for the 1982 book *Naming Names*

Following the hearing he resumed his career and worked with Elia Kazan and Budd Schulberg, two other HUAC "friendly witnesses", on the 1954 film *On the Waterfront*, which is widely seen as an allegory and apologia for testifying.

In 1957 he appeared in Sidney Lumet's *12 Angry Men* as the abrasive Juror #3. In 1959, on CBS' *DuPont Show of the Month*, he starred in the dual roles of Miguel de Cervantes and Don Quixote in the play *I, Don Quixote*, which years later became the musical *Man of La Mancha*. Cobb also appeared as Wyoming ranch owner Judge Henry Garth in the first four seasons of the long-running NBC western television series *The Virginian*. His co-stars were James Drury, Doug McClure, Roberta Shore, Gary Clarke, Randy Boone, Clu Gulager and Diane Roter.

In 1968 his performance as King Lear with Stacy Keach as Edmund, René Auberjonois as the Fool and Philip Bosco as Kent achieved the longest run (72 performances) for the play in Broadway history.[7]

One of his final film roles was that of police detective Lt. Kinderman in the 1973 horror film *The Exorcist*.

Cobb died of a heart attack in February 1976 in Woodland Hills, California, and was buried in Mount Sinai Memorial Park Cemetery in Los Angeles. He was survived by his second wife, Mary Hirsch.[8] His death came the day before his *Exodus* (1960) co-star Sal Mineo was murdered.

He was inducted, posthumously, into the American Theatre Hall of Fame in 1981.[9]

Lee J. Cobb's wife from 1940 to 1950s was Yiddish-theater and film actress Helen Beverley (1916—2011), daughter — Julie Cobb.[10] [11]

Selected Broadway credits

- *Crime and Punishment (1935)*
- *Waiting for Lefty* (1935)
- *Johnny Johnson* (1936)
- *Golden Boy* (1937)
- *Death of a Salesman* (1949)
- *King Lear* (1968)

Filmography

- *The Vanishing Shadow* (1934)
- *North of the Rio Grande* (1937)
- *Rustlers' Valley* (1937)
- *Ali Baba Goes to Town* (1937)
- *Danger on the Air* (1938)
- *The Phantom Creeps* (1939)
- *Golden Boy* (1939)
- *This Thing Called Love* (1940)
- *Men of Boys Town* (1941)
- *Paris Calling* (1941)
- *Down Rio Grande Way* (1942)
- *The Moon Is Down* (1943)
- *Tonight We Raid Calais* (1943)
- *Buckskin Frontier* (1943)
- *The Song of Bernadette* (1943)
- *Winged Victory* (1944)
- *Anna and the King of Siam* (1946)
- *Johnny O'Clock* (1947)
- *Boomerang!* (1947)
- *Captain from Castile* (1947)
- *Call Northside 777* (1948)
- *The Miracle of the Bells* (1948)
- *The Luck of the Irish* (1948)
- *The Dark Past* (1948)
- *Thieves' Highway* (1949)
- *The Man Who Cheated Himself* (1950)
- *Sirocco* (1951)
- *The Family Secret* (1951)
- *The Fighter* (1952)
- *The Tall Texan* (1953)
- *Yankee Pasha* (1954)
- *Gorilla at Large* (1954)
- *On the Waterfront* (1954)
- *Day of Triumph* (1954)
- *The Racers* (1955)
- *The Road to Denver* (1955)

Cobb as Johnny Friendly with Marlon Brando as *Terry Malloy* in *On the Waterfront* (1954)

Cobb as Johnny Friendly in *On the Waterfront* (1954)

- *The Left Hand of God* (1955)
- *The Man in the Gray Flannel Suit* (1956)
- *Miami Expose* (1956)
- *12 Angry Men* (1957)
- *The Garment Jungle* (1957)
- *The Three Faces of Eve* (1957)
- *The Brothers Karamazov* (1958)
- *Man of the West* (1958)
- *Party Girl* (1958)
- *The Trap* (1959)
- *Green Mansions* (1959)
- *But Not for Me* (1959)
- *Exodus* (1960)
- *The DuPont Show with June Allyson* as Captain Maximilian Gault in "School of the Soldier" (CBS, 1961)
- *The Final Hour* (1962)
- *Four Horsemen of the Apocalypse* (1962)
- *The Brazen Bell* (1962)
- *How the West Was Won* (1962)
- *Come Blow Your Horn* (1963)
- *Our Man Flint* (1966)
- *In Like Flint* (1967)
- *The Day of the Owl* (1968)
- *Coogan's Bluff* (1968)
- *They Came to Rob Las Vegas* (1968)
- *Mackenna's Gold* (1969)
- *The Liberation of L.B. Jones* (1970)
- *Macho Callahan* (1970)
- *Lawman* (1971)
- *The Man Who Loved Cat Dancing* (1973)
- *Ultimatum* (1973)
- *Ransom! Police Is Watching* (1973)
- *The Exorcist* (1973)
- *The Balloon Vendor* (1974)
- *Mark of the Cop* (1975)
- *That Lucky Touch* (1975)
- *Cross Shot* (1976)
- *Nick the Sting* (1976)

See also

- McCarthyism
- Second Red Scare

References

[1] Obituary *Variety*, February 18, 1976.
[2] Vernon Scott (January 4, 1976). "Bicentennial a 'very special event" for actor Lee J. Cobb" (http://news.google.com/newspapers?id=RMUdAAAAIBAJ&sjid=YIYEAAAAIBAJ&pg=6763,1901672&dq). *The Pittsburgh Press*. . Retrieved March 24, 2009.
[3] United States Census for 1920, Bronx (New York) Assembly District 4, District 254, Page 16
[4] "Lee J. Cobb Biography" (http://www.biography.com/articles/Lee-J.-Cobb-9542417). Biography.com. . Retrieved May 10, 2010.
[5] "Pinewood Lake website retrieved on 2010-09-10" (http://www.pinewoodlake.org/). Pinewoodlake.org. . Retrieved 2012-07-25.
[6] "World War II: The Movie | History of Flight | Air & Space Magazine" (http://www.airspacemag.com/history-of-flight/World-War-II-The-Movie.html?c=y&page=2). Airspacemag.com. . Retrieved 2012-07-25.
[7] The Broadway League. "King Lear | IBDB: The official source for Broadway Information" (http://www.ibdb.com/production.php?id=3432). IBDB. . Retrieved 2012-11-08.
[8] "Biography for Lee J. Cobb" (http://www.tcm.com/tcmdb/participant.jsp?spid=35965&apid=96791). tcm.com. . Retrieved May 12, 2010.
[9] ""The New York Times", March 3, 1981 - "26 Elected to the Theater Hall of Fame"" (http://www.nytimes.com/1981/03/03/theater/26-elected-theater-hall-fame-26-broadway-voted-into-theater-hall-fame.html). Nytimes.com. 1981-03-03. . Retrieved 2012-07-25.
[10] "Helen Beverley, Yiddish Theatre Actress, Dies at 94" (http://www.playbill.com/news/article/153051-Helen-Beverley-Yiddish-Theatre-Actress-Dies-at-94). Playbill.com. 2011-07-26. . Retrieved 2012-07-25.
[11] Helen Beverley, Star of Yiddish Films (http://www.jewishjournal.org/story/news_features/helen_beverley_star_of_yiddish_films/)

External links

- Lee J. Cobb (http://www.ibdb.com/person.asp?ID=67109) at the Internet Broadway Database
- Lee J. Cobb (http://www.imdb.com/name/nm2011/) at the Internet Movie Database
- Lee J. Cobb (http://www.findagrave.com/cgi-bin/fg.cgi?page=gr&GRid=4514) at *Find a Grave*

Darryl_F._Zanuck

Darryl F. Zanuck	
 Darryl F. Zanuck in his office, circa 1940.	
Born	Darryl Francis ZanuckSeptember 5, 1902Wahoo, Nebraska
Died	December 22, 1979 (aged 77)Palm Springs, California
Cause of death	Jaw cancer
Resting place	Westwood Village Memorial Park Cemetery
Other names	Gregory Rogers[1] Melville Crossman[1] Mark Canfield[1]
Years active	1922–1970
Spouse(s)	Virginia Fox (1924–1956; separated)
Children	Richard D. Zanuck Darrylin Zanuck Susan Zanuck

Darryl Francis Zanuck (September 5, 1902 – December 22, 1979) was an American producer, writer, actor, director and studio executive who played a major part in the Hollywood studio system as one of its longest survivors (the length of his career being rivaled only by that of Adolph Zukor). He earned three Academy Awards during his tenure.

Early life

Zanuck was born in Wahoo, Nebraska, the son of Louise (née Torpin) and Frank Zanuck, who owned and operated a hotel in Wahoo.[2] Zanuck was of part Swiss descent[2] and was raised a Protestant.[3] At six, Zanuck and his mother moved to Los Angeles, where the better climate could improve her poor health. At eight, he found his first movie job as an extra, but his disapproving father recalled him to Nebraska.

In 1918, despite being sixteen, he deceived a recruiter and joined the United States Army and served in France with the Nebraska National Guard. Returning to the US, he worked in many part-time jobs while seeking work as a writer. He found work producing movie plots, selling his first story in 1922 to William Russell and his second to Irving Thalberg. Screenwriter Frederica Sagor Maas, story editor at Universal Pictures' New York office, stated that one of the stories Zanuck sent out to movie studios around this time was completely plagiarized from another author's work.[4] Zanuck then worked for Mack Sennett and took that experience to Warner Brothers where he wrote stories for Rin Tin Tin and under a number of pseudonyms wrote over forty scripts from 1924 to 1929, including *Red Hot Tires* (1925) and *Old San Francisco* (1927). He moved into management in 1929, and became head of production in 1931.

Studio head

In 1933 he left Warners to found 20th Century Films with Joseph Schenck and William Goetz, releasing their material through United Artists. In 1935 Schenck and Zanuck bought out Fox studios to become 20th Century Fox. Zanuck was vice-president of this new studio and took an interventionist approach, closely involved in editing and producing. Like the other heads of Hollywood studios, during the war he was commissioned a Colonel in the Army Signal Corps. He returned to Fox in 1944.

In the 1950s, he withdrew from the studio to concentrate on independent producing in Europe. He left his wife, Virginia Fox, in 1956 and moved to Europe to concentrate on producing. Many of his later films were designed in part to promote the careers of his successive girlfriends, Bella Darvi, Irina Demick and Geneviève Gilles, and several movies he produced featured his girlfriend of moment, including the French singer Juliette Gréco.[5]

He returned to control Fox in 1962, replacing Spyros Skouras, in a confrontation over the release of Zanuck's production of *The Longest Day* as the studio struggled to finish the difficult production of *Cleopatra*. He made his son Richard D. Zanuck head of production. He became involved in a power struggle with the board and his son circa 1969. In May 1971 Zanuck was forced from "his" studio.

Death

A long time cigar smoker,[6] he died of jaw cancer at the age of 77, and was interred in the Westwood Village Memorial Park Cemetery.

Darryl Zanuck's grave

Legacy

For his contribution to the motion picture industry, Darryl F. Zanuck has a star on the Hollywood Walk of Fame at 6336 Hollywood Blvd and has won 3 Thalberg Awards of the Academy of Motion Picture Arts and Sciences. On the present-day Fox lot, movies are shown in the Zanuck Theater.

Zanuck is a minor character in the fictional musical *Bombshell* in the first season of the TV series *Smash* and sings "Don't Say Yes Until I Finish Talking" and "Smash".

Academy Awards

Year	Result	Category	Film
1929–30	Nominated	Outstanding Production	*Disraeli*
1932–33	Nominated	Outstanding Production	*42nd Street*
1934	Nominated	Outstanding Production	*The House of Rothschild*
1935	Nominated	Outstanding Production	*Les Misérables*
1937	Nominated	Outstanding Production	*In Old Chicago*
1938	Nominated	Outstanding Production	*Alexander's Ragtime Band*
1940	Nominated	Outstanding Production	*The Grapes of Wrath*
1941	**Won**	Outstanding Motion Picture	*How Green Was My Valley*
1944	Nominated	Outstanding Motion Picture	*Wilson*
1946	Nominated	Outstanding Motion Picture	*The Razor's Edge*
1947	**Won**	Outstanding Motion Picture	*Gentleman's Agreement*
1949	Nominated	Outstanding Motion Picture	*Twelve O'Clock High*
1950	**Won**	Outstanding Motion Picture	*All About Eve*
1962	Nominated	Best Picture	*The Longest Day*

Quotes

- On discovering actress Gene Tierney after appearing on Broadway in The Male Animal after her first film "undeniably the most beautiful actress in movie history."
- In 1946 Zanuck said "(Television) won't be able to hold on to any market it captures after the first six months. People will soon get tired of staring at a plywood box every night."[7]

References

[1] Per IMDb.
[2] http://www.wahooschools.org/vnews/display.v/SEC/Wahoo's%20Famous%20Sons%3E%3EDarryl%20Zanuck
[3] Gussow, Mel (September 1, 2002). "FILM; Darryl F. Zanuck, Action Hero of the Studio Era" (http://www.nytimes.com/2002/09/01/movies/film-darryl-f-zanuck-action-hero-of-the-studio-era.html?pagewanted=1). *The New York Times*. . Retrieved May 1, 2010.
[4] Maas, Frederica Sagor (1999). *The Shocking Miss Pilgrim: A Writer in Early Hollywood*. Lexington, KY: The University Press of Kentucky. pp. 44-45. ISBN 0-8131-2122-1.
[5] Charlotte Mosley, editor, *In Tearing Haste: Letters Between Deborah Devonshire and Patrick Leigh-Fermor*, 2008, John Murray
[6] Hift, Fred (1 September 1994). "The Longest Day" (http://www.cigaraficionado.com/webfeatures/show/id/The-Longest-Day_6102). *Cigar Aficionado*. . Retrieved 6 December 2011.
[7] Biography for Darryl F. Zanuck (http://www.imdb.com/name/nm953123/bio) at the Internet Movie Database

Further reading

- Behlmer, Rudy (editor) (1993). *Memo from Darryl F. Zanuck: The Golden Years at Twentieth Century-Fox.* Grove. ISBN 0-8021-1540-3.
- Mosley, Leonard (1984). *Zanuck: The Rise and Fall of Hollywood's Last Tycoon*. Little, Brown. ISBN 0-316-58538-6.
- Thackrey Jr., Thomas. (December 23, 1979). "Darryl F. Zanuck, Last of Movie Moguls, Dies at 77". *Los Angeles Times*, p. 1.

External links

- Darryl F. Zanuck (http://www.imdb.com/name/nm0953123/) at the Internet Movie Database
- *The Zanucks: Reel Royalty* (http://www.cbsnews.com/stories/2005/07/08/sunday/main707725.shtml) from *CBS News Sunday Morning*, July 10, 2005. Accessed November 30, 2006.

Winged Victory (film)

WARNING: Article could not be rendered - ouputting plain text.

Potential causes of the problem are: (a) a bug in the pdf-writer software (b) problematic Mediawiki markup (c) table is too wide

Winged VictoryDirected byGeorge CukorProduced byDarryl F. ZanuckWritten byMoss Hart, based upon his playStarringLon McCallisterJeanne CrainRed ButtonsDon Taylor (actor)Don TaylorDistributed by20th Century Fox in association with the U.S. Army Air ForcesRelease date(s) 22 December 1944Country United StatesLanguage EnglishWinged Victory is a 1944 in film1944 drama film directed by George Cukor, a joint effort of 20th Century Fox and the U.S. Army Air Forces. Based upon the successful Winged Victory (play)play with the same name by Moss Hart, who also wrote the screenplay, the film only opened after the play's theatre run.Plot Frankie Davis (Lon McCallister), Allan Ross (Mark Daniels) and Pinkie Scariano (Don Taylor) join the U.S. Army Air Forces with hopes of becoming pilots. In training, they befriend Irving Miller (Edmond O'Brien) and Bobby Grills (Barry Nelson). The five friends go through the training process to become pilots, facing success, failure and tragedy. CastLon McCallister - Frankie Davis Jeanne Crain - Helen Edmond O'Brien - Irving Miller Jane Ball - Jane Preston Mark Daniels - Alan Ross Jo-Carroll Dennison - Dorothy Ross Don Taylor (actor)Don Taylor - Danny 'Pinkie' Scariano Judy Holliday - Ruth Miller Lee J. Cobb - Doctor Peter Lind Hayes - O'Brien Red Buttons - Whitey / Andrews Sister Barry Nelson - Bobby Crills Rune Hultman - Dave Anderson Gary Merrill - Capt. McIntyre George Reeves - Lt. Thompson External links Winged Victory at the Internet Movie Database

George_Cukor

George Cukor	
 Cukor in 1973	
Born	George Dewey CukorJuly 7, 1899New York City, New York, U.S.
Died	January 24, 1983 (aged 83)Los Angeles, California, U.S.
Resting place	Forest Lawn Memorial Park, Glendale, California
Occupation	Director
Years active	1930–81

George Dewey Cukor (pron.: /ˈkjuːkər/; July 7, 1899 – January 24, 1983) was an American film director.[1] He mainly concentrated on comedies and literary adaptations. His career flourished at RKO when David O. Selznick, the studio's Head of Production, assigned Cukor to direct several of RKO's major films including *What Price Hollywood?* (1932), *A Bill of Divorcement* (1932), "Our Betters" (1933), and *Little Women* (1933). When Selznick moved to MGM in 1933 Cukor followed and directed *Dinner at Eight* (1933) and *David Copperfield* (1935) for Selznick and *Romeo and Juliet* (1936) and *Camille* (1936) for Irving Thalberg.

He was replaced as the director of *Gone with the Wind* (1939), but he went on to direct *The Philadelphia Story* (1940), *Adam's Rib* (1949), *Born Yesterday* (1950), *A Star Is Born* (1954) and *My Fair Lady* (1964). He continued to work into the 1980s.

Early life

Cukor was born on the Lower East Side of New York City, the younger child and only son of Hungarian Jewish immigrants Victor, an assistant district attorney, and Helen Ilona (née Gross) Cukor. His parents selected his middle name in honor of Spanish–American War hero George Dewey. The family was not particularly religious, pork was a staple on the dinner table, and when he started attending temple as a boy, Cukor learned Hebrew phonetically, with no real understanding of the meaning of the words or what they represented. As a result, he was ambivalent about his faith and dismissive of old world traditions from childhood, and as an adult he embraced Anglophilia to remove himself even further from his roots.[2]

As a child, Cukor appeared in several amateur plays and took dance lessons, and at the age of seven he performed in a recital with David O. Selznick, who in later years would become a mentor and friend.[3] As a teenager, Cukor frequently was taken to the New York Hippodrome by his uncle. Infatuated with theatre, he often cut classes at DeWitt Clinton High School to attend afternoon matinees.[4] [5] During his senior year, he worked as a supernumerary with the Metropolitan Opera, earning 50¢ per appearance, and $1 if he was required to perform in blackface.[6]

Following his graduation in 1917, Cukor was expected to follow in his father's footsteps and pursue a career in law. He halfheartedly enrolled in the City College of New York, where he entered the Students Army Training Corps in October 1918. His military experience was limited; Germany surrendered in early November, and Cukor's duty ended after only two months. Shortly after, he left school.[7]

Cukor obtained a job as an assistant stage manager and bit player with a touring production of *The Better 'Ole*, a popular British musical based on Old Bill, a cartoon character created by Bruce Bairnsfather.[8] In 1920, he became the stage manager for the Knickerbocker Players, a troupe that shuttled between Syracuse and Rochester, New York, and the following year he was hired as general manager of the newly formed Lyceum Players, an upstate summer stock company. In 1925 he formed the C.F. and Z. Production Company with Walter Folmer and John Zwicki, which gave him his first opportunity to direct.[9] [10] Following their first season, he made his Broadway directorial debut with *Antonia* by Hungarian playwright Melchior Lengyel, then returned to Rochester, where C.F. and Z. evolved into the Cukor-Kondolf Stock Company, a troupe that included Louis Calhern, Ilka Chase, Phyllis Povah, Frank Morgan, Reginald Owen, Elizabeth Patterson and Douglass Montgomery, all of whom would work with Cukor in later years in Hollywood.[11] Lasting only one season with the company was Bette Davis. Cukor later recalled, "Her talent was apparent, but she did buck at direction. She had her own ideas, and though she only did bits and ingenue roles, she didn't hesitate to express them." For the next several decades, Davis claimed she was fired, and although Cukor never understood why she placed so much importance on an incident he considered so minor, he never worked with her again.[12]

For the next few years, Cukor alternated between Rochester in the summer months and Broadway in the winter. His direction of a 1926 stage adaptation of *The Great Gatsby* by Owen Davis brought him to the attention of the New York critics. Writing in the *Brooklyn Eagle*, drama critic Arthur Pollock called it "an unusual piece of work by a director not nearly so well-known as he should be."[13] Cukor directed six more Broadway productions before departing for Hollywood in 1929.

Early Hollywood career

When Hollywood began to recruit New York theater talent for sound films, Cukor immediately answered the call. In December 1928, Paramount Pictures signed him to a contract that reimbursed him for his airfare and initially paid him $600 per week with no screen credit during a six-month apprenticeship. He arrived in Hollywood in February 1929, and his first assignment was to coach the cast of *River of Romance* to speak with an acceptable Southern accent.[14] In October, the studio lent him to Universal Pictures to conduct the screen tests and work as a dialogue director for *All Quiet on the Western Front*. In 1930, he co-directed three films at Paramount, and his weekly salary was increased to $1500.[15] In 1931, he made his solo directorial debut with *Tarnished Lady* starring Tallulah Bankhead.

Cukor was then assigned to *One Hour with You* (1932), an operetta with Maurice Chevalier and Jeanette MacDonald, when original director Ernst Lubitsch opted to concentrate on producing the film instead. At first the two men worked well together, but two weeks into filming Lubitsch began arriving on the set on a regular basis, and he soon began directing scenes with Cukor's consent. Upon the film's completion, Lubitsch approached Paramount general manager B. P. Schulberg and threatened to leave the studio if Cukor's name wasn't removed from the credits. When Schulberg asked him to cooperate, Cukor filed suit. He eventually settled for being billed as assistant director and then left Paramount to work with David O. Selznick at RKO Studios.[16]

Cukor quickly earned a reputation as a director who could coax great performances from actresses and he became known as a 'woman's director', a title he resented. Despite this reputation, during his career, he oversaw more performances honored with the Academy Award for Best Actor than any other director: James Stewart in *The Philadelphia Story* (1940), Ronald Colman in *A Double Life* (1947), and Rex Harrison in *My Fair Lady* (1964). One of Cukor's first ingenues was actress Katharine Hepburn, who debuted in *A Bill of Divorcement* and whose looks and personality left RKO officials at a loss as to how to use her. Cukor directed her in several films, both successful

(*Little Women*, 1933) and disastrous (*Sylvia Scarlett*, 1935), and they became close friends off the set.

Cukor was hired to direct *Gone with the Wind* by Selznick in 1936, even before the book was published.[17] He spent the next two years involved with pre-production duties, including supervision of the numerous screen tests of actresses anxious to portray Scarlett O'Hara. Cukor favored Hepburn for the role, but Selznick, concerned about her reputation as 'box office poison', would not consider her without a screen test, and the actress refused to film one. Of those who did, Cukor preferred Paulette Goddard, but her supposedly illicit relationship with Charlie Chaplin (they were, in fact, secretly married) concerned Selznick.[18]

Between his *Wind* chores, the director assisted with other projects. He filmed the cave scene for *The Adventures of Tom Sawyer* (1938)[19] and, following the firing of its original director Richard Thorpe, Cukor spent a week on the set of *The Wizard of Oz* (1939). Although he filmed no footage, he made crucial changes to the look of Dorothy by eliminating Judy Garland's blonde wig and adjusting her makeup and costume, encouraging her to act in a more natural manner that greatly contributed to the success of the final film.[20] [21] Additionally, Cukor softened the Scarecrow's makeup and gave Margaret Hamilton's Wicked Witch of the West a different hairstyle, as well as altering her makeup and other facial features.

Cukor spent many hours coaching Vivien Leigh and Olivia de Havilland prior to the start of filming *Wind*, but Clark Gable resisted his efforts to get him to master a Southern accent. However, despite rumors about Gable being uncomfortable with Cukor on the set, nothing in the internal memos of David O. Selznick indicates or suggests that Clark Gable played any role in Cukor's dismissal from the film. Rather, they show Selznick's mounting dissatisfaction with Cukor's slow pace and quality of work. From a private letter from journalist Susan Myrick to Margaret Mitchell in February 1939: "George [Cukor] finally told me all about it. He hated [leaving the production] very much he said but he could not do otherwise. In effect he said he is an honest craftsman and he cannot do a job unless he knows it is a good job and he feels the present job is not right. For days, he told me he has looked at the rushes and felt he was failing... the things did not click as it should. Gradually he became convinced that the script was the trouble... So George just told David he would not work any longer if the script was not better and he wanted the [Sidney] Howard script back... he would not let his name go out over a lousy picture... And bull-headed David said 'OK get out!'"[22]

Selznick had already been unhappy with Cukor ("a very expensive luxury") for not being more receptive to directing other Selznick assignments, even though Cukor had remained on salary since early 1937; and in a confidential memo written in September 1938, four months before principal photography began, Selznick flirted with the idea of replacing him with Victor Fleming. "I think the biggest black mark against our management to date is the Cukor situation and we can no longer be sentimental about it.... We are a business concern and not patrons of the arts..." Cukor was relieved of his duties, but he continued to work with Leigh and De Havilland off the set. Various rumors about the reasons behind his dismissal circulated throughout Hollywood. Selznick's friendship with Cukor had crumbled slightly when the director refused other assignments, including *A Star is Born* (1937) and *Intermezzo* (1939).[23] Given that Gable and Cukor had worked together before, in *Manhattan Melodrama* and Gable had no objection to working with him then, and given Selznick's desperation to get Gable for Rhett Butler, if Gable had any objections to Cukor, certainly they would have been expressed before he signed his contract for the film.[24] Yet, writer Gore Vidal, in his autobiography *Point to Point Navigation*, recounted that Gable demanded that Cukor be fired off *Wind* because, according to Cukor, the young Gable had been a male hustler and Cukor had been one of his johns.[25] This has been confirmed by Hollywood biographer E. J. Fleming, who has recounted that, during a particularly difficult scene, Gable erupted publicly, screaming: "I can't go on with this picture. I won't be directed by a fairy. I have to work with a real man."[26]

Cukor's dismissal from *Wind* freed him to direct *The Women* (1939), notable for its all-female cast, followed by *The Philadelphia Story* (1940), starring Katharine Hepburn. He also directed another of his favorite actresses, Greta Garbo, in *Two-Faced Woman* (1941), her last film before she retired from the screen.

In 1942, at the age of forty-three, Cukor enlisted in the Signal Corps. Following basic training at Fort Monmouth, he was assigned to the old Paramount studios in Astoria, Queens (where he had directed three films in the early 1930s), although he was permitted to lodge at the St. Regis Hotel in Manhattan. Working with Irwin Shaw, John Cheever and William Saroyan, among others, Cukor produced training and instructional films for army personnel. Because he lacked an officer's commission, he found it difficult to give orders and directions to his superiors. Despite his efforts to rise above the rank of private—he even called upon Frank Capra to intercede on his behalf—he never achieved officer's status or any commendations during his six months of service. In later years Cukor suspected his homosexuality impeded him from receiving any advances or honors,[27] although rumors to that effect could not be confirmed.[28]

The remainder of the decade was a series of hits and misses for Cukor. Both *Two Faced Woman* and *Her Cardboard Lover* (1942) were commercial failures. More successful were *A Woman's Face* (1941) with Joan Crawford and *Gaslight* (1944) with Ingrid Bergman and Charles Boyer. During this era, Cukor forged an alliance with screenwriters Garson Kanin and Ruth Gordon, who had met in Cukor's home in 1939 and married three years later. Over the course of seven years, the trio collaborated on seven films, including *A Double Life (1947) starring Ronald Colman, Adam's Rib* (1949), *Born Yesterday* (1950), *The Marrying Kind* (1952), and *It Should Happen to You* (1954), all starring another Cukor favorite, Judy Holliday, who won the Academy Award for Best Actress for *Born Yesterday.*

Later Hollywood career

In December 1952, Cukor was approached by Sid Luft, who proposed the director helm a musical remake of the 1937 film *A Star is Born* with his then-wife Judy Garland in the lead role. Cukor had declined to direct the earlier film because it was too similar to his 1932 *What Price Hollywood?*, but the opportunity to direct his first Technicolor film, first musical, and work with screenwriter Moss Hart and especially Garland appealed to him, and he accepted.[29] Getting the updated *A Star Is Born* to the screen proved to be a challenge. Cukor wanted Cary Grant for the male lead and went so far as to read the entire script with him, but Grant, while agreeing it was the role of a lifetime, steadfastly refused to do it, and Cukor never forgave him. The director then suggested either Humphrey Bogart or Frank Sinatra tackle the part, but Jack Warner rejected both. Stewart Granger was the front runner for a period of time, but he backed out when he was unable to adjust to Cukor's habit of acting out scenes as a form of direction.[30] James Mason ultimately was signed, and filming began on October 12, 1953. As the months passed, Cukor was forced to deal not only with constant script changes but a very unstable leading lady, who was plagued by chemical and alcohol dependencies, extreme weight fluctuations, and real and imagined illnesses. In March 1954, a rough cut still missing several musical numbers was assembled, and Cukor had mixed feelings about it. When the last scene finally was filmed in the early morning hours of July 28, 1954, Cukor already had departed the production and was unwinding in Europe.[31] The first preview the following month ran 210 minutes and, despite ecstatic feedback from the audience, Cukor and editor Folmar Blangsted trimmed it to 182 minutes for its New York premiere in October. The reviews were the best of Cukor's career, but Warner executives, concerned the running time would limit the number of daily showings, made drastic cuts without Cukor, who had departed for India to scout locations for *Bhowani Junction.* At its final running time of 154 minutes, the film had lost musical numbers and crucial dramatic scenes, and Cukor called it "very painful."[32] He was not included in the film's six Oscar nominations, all of which were lost.

Over the next ten years, Cukor directed a handful of films with varying success. *Les Girls* (1957) won the Golden Globe Award for Best Motion Picture – Musical or Comedy, and *Wild Is the Wind* (1957) earned Oscar nominations for Anna Magnani and Anthony Quinn, but neither *Heller in Pink Tights* nor *Let's Make Love* (both 1960) were box office hits. His most notable project during this period was the ill-fated *Something's Got to Give*, an updated remake of the 1940 screwball comedy *My Favorite Wife*. Cukor liked leading lady Marilyn Monroe but found it difficult to deal with her erratic work habits, frequent absences from the set, and the constant presence of her acting coach, Paula Strasberg. After thirty-two days of shooting, the director had only 7½ minutes of usable film.[33] Then Monroe travelled to New York to appear at a birthday celebration for John F. Kennedy at Madison Square Garden, where she famously serenaded the President. Studio documents released after Monroe's death confirmed that her appearance at the political fundraising event was approved by Fox executives. The production came to a halt when Cukor had filmed every scene not involving Monroe and the actress remained unavailable. 20th Century Fox executive Peter Levathes fired her and hired Lee Remick to replace her, prompting co-star Dean Martin to quit, since his contract guaranteed he would be playing opposite Monroe.[34] With the production already $2 million over budget[33] and everyone back at the starting gate, the studio pulled the plug on the project. Less than two months later, Monroe was found dead in her home.

George Cukor reading in his house, by Allan Warren

Two years later, Cukor achieved one of his greatest successes with *My Fair Lady*. Throughout filming there were mounting tensions between the director and designer Cecil Beaton, but Cukor was thrilled with leading lady Audrey Hepburn, although the crew was less enchanted with her diva-like demands.[35] Although several reviews were critical of the film – Pauline Kael said it "staggers along" and Stanley Kauffmann thought Cukor's direction was like "a rich gravy poured over everything, not remotely as delicately rich as in the Asquith-Howard 1937 *Pygmalion*" —[36] the film was a box office hit which won him the Academy Award for Best Director, the Golden Globe Award for Best Director, and the Directors Guild of America Award after having been nominated for each several times.

Following *My Fair Lady*, Cukor became less active. He directed Maggie Smith in *Travels with My Aunt* in 1972 and helmed the critical and commercial flop *The Blue Bird*, the first joint Soviet-American production, in 1976. He reunited twice with Katharine Hepburn for the television movies *Love Among the Ruins* (1975) and *The Corn Is Green* (1979). He directed his final film, *Rich and Famous* (1981) with Jacqueline Bisset and Candice Bergen, at the age of eighty-two.

In 1976, Cukor was awarded the George Eastman Award, given by George Eastman House for distinguished contribution to the art of film.[37]

Personal life

George Cukor in his garden designed by Florence Yoch & Lucile Council, Hollywood Hills, Los Angeles

It was an open secret in Hollywood that Cukor was homosexual, although he was discreet about his sexual orientation and "never carried it as a pin on his lapel", as producer Joseph L. Mankiewicz put it.[38] He was a celebrated *bon vivant* whose luxurious home was the site of weekly Sunday afternoon parties attended by closeted celebrities and the attractive young men they met in bars and gyms and brought with them.[39] At least once, in the midst of his reign at MGM, he was arrested on vice charges, but studio executives managed to get the charges dropped and all records of it expunged, and the incident never was publicized by the press.[40] In the late 1950s, Cukor became involved with a considerably younger man named George Towers. He financed his education at the Los Angeles State College of Applied Arts and Sciences and the University of Southern California, from which Towers graduated with a law degree in 1967.[41] That fall Towers married and his relationship with Cukor evolved into one of father and son, and for the remainder of Cukor's life the two remained very close.[42]

By the mid-1930s, Cukor was not only established as a prominent director but, socially, as an unofficial head of Hollywood's gay subculture. His home, redecorated in 1935 by actor-turned interior designer William Haines with gardens designed by Florence Yoch & Lucile Council, was the scene of many gatherings for the industry's homosexuals. The close-knit group included Haines and his partner Jimmie Shields, Alan Ladd, writer Somerset Maugham, director James Vincent, screenwriter Rowland Leigh, costume designers Orry-Kelly and Robert Le Maire, and actors John Darrow, Anderson Lawler, Grady Sutton, Robert Seiter and Tom Douglas. Frank Horn, secretary to Cary Grant, was also a frequent guest.[43]

Cukor's friends were of paramount importance to him and he kept his home filled with their photographs. Regular attendees at his famed soirées included Katharine Hepburn and Spencer Tracy, Joan Crawford and Douglas Fairbanks, Jr., Lauren Bacall and Humphrey Bogart, Claudette Colbert, Marlene Dietrich, Laurence Olivier and Vivien Leigh, actor Richard Cromwell, Stanley Holloway, Judy Garland, Gene Tierney, Noël Coward, Cole Porter, director James Whale, costume designer Edith Head, and Norma Shearer, especially after the death of her first husband, Irving Thalberg. He often entertained literary figures like Sinclair Lewis, Theodore Dreiser, Aldous Huxley, Ferenc Molnár, and close friend Somerset Maugham, as well.[44]

The PBS series American Masters produced a comprehensive documentary about his life and work titled *On Cukor* directed by Robert Trachtenberg in 2000.

Death

Cukor died of a heart attack on January 24, 1983, and was interred in an unmarked grave at the Forest Lawn Memorial Park Cemetery in Glendale, California.[45] Records in probate court indicated his net worth at the time of his death was $2,377,720.[46]

Filmography

Year	Title	Studio	Genre	Cast	Notes
1930	*Grumpy*	Paramount Pictures	Drama	Cyril Maude	Co-directed with Cyril Gardner
1930	*The Virtuous Sin*	Paramount Pictures	Drama	Kay Francis, Walter Huston, Kenneth MacKenna	Co-directed with Louis J. Gasnier
1930	*The Royal Family of Broadway*	Paramount Pictures	Comedy	Fredric March, Ina Claire	Co-directed with Cyril Gardner
1931	*Tarnished Lady*	Paramount Pictures	Drama	Tallulah Bankhead, Clive Brook, Alexander Kirkland	
1931	*Girls About Town*	Paramount Pictures	Comedy	Kay Francis, Lilyan Tashman, Joel McCrea	
1932	*One Hour with You*	Paramount Pictures	Musical	Maurice Chevalier, Jeanette MacDonald	Directed part of the film when Ernst Lubitsch took ill and was credited as dialogue director
1932	*Une heure près de toi*	Paramount Pictures	Musical	Maurice Chevalier, Jeanette MacDonald	French version of One Hour With You / Co-directed with Ernst Lubitsch
1932	*What Price Hollywood?*	RKO	Drama	Constance Bennett, Lowell Sherman, Neil Hamilton	
1932	*A Bill of Divorcement*	RKO	Drama	John Barrymore, Billie Burke, Katharine Hepburn	
1932	*Rockabye*	RKO	Drama	Constance Bennett, Joel McCrea, Paul Lukas	David O. Selznick hired Cukor to film two weeks of retakes after the initial film by George Fitzmaurice did not meet expectations
1932	*The Animal Kingdom*	RKO	Drama	Leslie Howard, Ann Harding, Myrna Loy	Uncredited
1933	*Our Betters*	RKO	Drama	Constance Bennett	

1933	*Dinner At Eight*	MGM	Drama	John Barrymore, Jean Harlow, Marie Dressler, Lionel Barrymore, Billie Burke, Wallace Beery	
1933	*Little Women*	RKO	Drama	Katharine Hepburn, Joan Bennett, Douglass Montgomery	
1934	*Manhattan Melodrama*	MGM, Cosmopolitan Pictures	Crime	Clark Gable, William Powell, Myrna Loy	Directed additional scenes after production
1935	*David Copperfield*	MGM	Drama	Freddie Bartholomew, W. C. Fields, Lionel Barrymore	
1935	*No More Ladies*	MGM	Comedy	Joan Crawford, Robert Montgomery, Franchot Tone	Completed filming when Edward H. Griffith took ill
1935	*Sylvia Scarlett*	RKO	Comedy	Katharine Hepburn, Cary Grant, Brian Aherne	
1936	*Romeo and Juliet*	MGM	Romance	Norma Shearer, Leslie Howard, John Barrymore, Basil Rathbone	
1936	*Camille*	MGM	Romance	Greta Garbo, Robert Taylor, Henry Daniell	
1938	*I Met My Love Again*	Walter Wanger Productions	Romance	Joan Bennett, Henry Fonda	Assisted Joshua Logan in directing parts of the film
1938	*The Adventures of Tom Sawyer*	David O. Selznick	Adventure	Tommy Kelly, Jackie Moran	Shot some retakes after production completed
1938	*Holiday*	Columbia	Drama	Katharine Hepburn, Cary Grant	
1939	*Zaza*	Paramount	Drama	Claudette Colbert, Herbert Marshall	
1939	*The Women*	MGM	Drama	Norma Shearer, Joan Crawford, Rosalind Russell	
1939	*Gone With the Wind*	David O. Selznick	Drama	Vivien Leigh, Clark Gable, Olivia De Havilland, Leslie Howard	Fired in the early stages of production, a few of his scenes remain in the finished film
1940	*Susan and God*	MGM	Comedy	Joan Crawford, Fredric March	
1940	*The Philadelphia Story*	MGM	Comedy	Katharine Hepburn, Cary Grant, Jimmy Stewart	

1941	*A Woman's Face*	MGM	Drama	Joan Crawford, Melvyn Douglas, Conrad Veidt	
1941	*Two-Faced Woman*	MGM	Comedy	Greta Garbo, Melvyn Douglas, Constance Bennett	
1942	*Her Cardboard Lover*	MGM	Comedy	Norma Shearer, Robert Taylor, George Sanders	
1942	*Keeper of the Flame*	MGM	Drama	Spencer Tracy, Katharine Hepburn	
1943	*Resistance and Ohm's Law*	Army Signal Corps	Documentary		short film
1944	*Gaslight*	MGM	Drama	Ingrid Bergman, Charles Boyer, Joseph Cotten	
1944	*Winged Victory*	20th Century Fox	War	Ensemble cast	
1944	*I'll Be Seeing You*	David O. Selznick	Drama	Ginger Rogers, Joseph Cotten, Shirley Temple	Replaced by William Dieterle during production
1947	*Desire Me*	MGM	Drama	Greer Garson, Robert Mitchum	Contributed to the film along with 4 other directors
1947	*A Double Life*	Garson Kanin	Film noir	Ronald Colman, Signe Hasso, Shelley Winters	
1949	*Edward, My Son*	MGM	Drama	Spencer Tracy, Deborah Kerr	
1949	*Adam's Rib*	MGM	Comedy	Spencer Tracy, Katharine Hepburn, Judy Holliday	
1950	*A Life of Her Own*	MGM	Drama	Lana Turner, Ray Milland	
1950	*Born Yesterday*	Columbia	Comedy	Judy Holliday, Broderick Crawford, William Holden	
1951	*The Model and the Marriage Broker*	20th Century Fox	Comedy	Thelma Ritter, Jeanne Crain, Scott Brady	
1952	*The Marrying Kind*	Columbia	Comedy	Judy Holliday, Aldo Ray	
1952	*Pat and Mike*	MGM	Comedy	Spencer Tracy, Katharine Hepburn, Aldo Ray	

1953	*The Actress*	MGM	Comedy	Spencer Tracy, Teresa Wright, Jean Simmons	
1954	*It Should Happen to You*	Columbia	Comedy	Judy Holliday, Peter Lawford, Jack Lemmon	
1954	*A Star Is Born*	Warner Bros., Transcona Enterprises	Drama	Judy Garland, James Mason	Partially lost film
1956	*Bhowani Junction*	MGM	Drama	Ava Gardner, Stewart Granger	
1957	*Les Girls*	MGM	Musical	Gene Kelly, Mitzi Gaynor, Kay Kendall	
1957	*Wild Is the Wind*	Paramount Pictures	Drama	Anna Magnani, Anthony Quinn	
1958	*Hot Spell*	Paramount Pictures	Drama	Shirley Booth, Anthony Quinn, Shirley MacLaine	Uncredited
1960	*Heller in Pink Tights*	Paramount Pictures	Western	Sophia Loren, Anthony Quinn, Steve Forrest	The final film was disavowed by Cukor
1960	*Song Without End*	William Goetz	Drama	Dirk Bogarde, Capucine, Genevieve Page	Completed the film when Charles Vidor died during production
1960	*Let's Make Love*	The Company of Artists	Musical	Marilyn Monroe, Yves Montand, Tony Randall	
1962	*Something's Got to Give*	20th Century Fox	Comedy	Marilyn Monroe, Dean Martin, Cyd Charisse	The film was abandoned after Monroe's death / 37 minutes of footage survives
1962	*The Chapman Report*	DFZ Productions	Drama	Shelley Winters, Jane Fonda, Claire Bloom, Glynis Johns	
1964	*My Fair Lady*	Warner Bros.	Musical	Audrey Hepburn, Rex Harrison	
1969	*Justine*	20th Century Fox	Drama	Michael York, Anouk Aimée, Dirk Bogarde	Replaced Joseph Strick shortly after production began
1972	*Travels with My Aunt*	MGM	Comedy	Maggie Smith, Alec McCowen, Cindy Williams	
1976	*The Blue Bird*	20th Century Fox, Lenfilm Studio, Tower International, Wenks Films	Drama	Elizabeth Taylor, Jane Fonda, Ava Gardner	
1981	*Rich And Famous*	MGM, Jaquet	Drama	Jacqueline Bisset, Candice Bergen	

- *Love Among the Ruins* (1975)
- *The Corn is Green* (1979)

References

[1] Obituary *Variety*, January 26, 1983.
[2] McGilligan, Patrick, *George Cukor: A Double Life*. New York: St. Martin's Press 1991. ISBN 0-312-05419-X, pp. 5–6.
[3] McGilligan, p. 11.
[4] Kipen, David. "Flawed look at career of blacklisted director" (http://www.sfgate.com/cgi-bin/article.cgi?f=/c/a/2001/08/29/DD141675.DTL), *San Francisco Chronicle*, August 29, 2001. Accessed September 14, 2009. "The American 20th century went to high school at DeWitt Clinton High in the Bronx. Multicultural before there was a name for it – at least a polite one --Clinton nurtured such diverse and influential figures as Bill Graham, James Baldwin, George Cukor, Neil Simon and Abraham Lincoln Polonsky."
[5] McGilligan, p. 10.
[6] Levy, Emanuel, *George Cukor: Master of Elegance*. New York: William Morrow & Company, Inc. 1994. ISBN 0-688-11246-3, pp. 26–27.
[7] McGilligan, p. 19.
[8] McGilligan, p. 21.
[9] Levy, pp. 33–34.
[10] McGilligan, pp. 34–35.
[11] McGilligan, pp. 36–41.
[12] Levy, pp. 36–37.
[13] McGilligan, p. 53.
[14] McGilligan, p. 61.
[15] McGilligan, pp. 67–69.
[16] McGilligan, pp. 69–71.
[17] McGilligan, p. 134.
[18] McGilligan, pp. 137–38.
[19] McGilligan, pp. 139–40.
[20] "*The Wizard of Oz* at Turner Classic Movies" (http://www.tcm.com/tcmdb/title.jsp?stid=852&category=Notes). Tcm.com. . Retrieved 2010-05-03.
[21] McGilligan, p. 145.
[22] Myrick, Susan White (1986), *Columns in Hollywood: Reports from the Gwtw Sets*, Mercer University Press.
[23] McGilligan, p. 139.
[24] *Hollywood Studio Magazine*, "The Great Directors" September 1986.
[25] Vidal, Gore (2007). *Point to Point Navigation: A Memoir* (http://books.google.com/books?id=yP8KHNtT69gC&printsec=frontcover#v=onepage&q&f=false). New York: Random House. pp. 135. ISBN 978-0-307-27501-1. .
[26] Fleming, E. J. (2005). *The Fixers: Eddie Mannix, Howard Strickling, and the MGM Publicity Machine* (http://books.google.com/books?id=f0usSUISUUMC&printsec=frontcover#v=onepage&q&f=false). Jefferson NC: McFarland. pp. 182. ISBN 978-0-7864-2027-8. .
[27] McGilligan, pp. 171–75.
[28] Levy, p. 150.
[29] McGilligan, 217-18.
[30] McGilligan, pp. 219–20.
[31] McGilligan, pp. 224–26.
[32] McGilligan, pp. 236–37.
[33] Levy, p. 271.
[34] McGilligan, p. 272.
[35] Levy, p. 289.
[36] Levy, p. 293.
[37] The George Eastman Award (http://www.eastmanhouse.org/museum/awards.php)
[38] McGilligan, p. 113.
[39] McGilligan, pp. 186–87.
[40] McGilligan, p. 133.
[41] McGilligan, pp. 277–78.
[42] McGilligan, pp. 307, 347–48.
[43] Mann, William J.; *Wisecracker: The Life and Times of William Haines, Hollywood's first Openly Gay Star*; New York: Viking, 1998; pp. 253, 255, 256.
[44] McGilligan, pp. 124–25.
[45] "Findagrave.com" (http://www.findagrave.com/cgi-bin/fg.cgi?page=gr&GRid=3718). Findagrave.com. . Retrieved 2010-05-03.
[46] McGilligan, p. 343.

Bibliography

- Hillstrom, Laurie Collier, *International Dictionary of Films and Filmmakers*. Detroit: St. James Press, 1997. ISBN 1-55862-302-7.
- Katz, Ephraim, *The Film Encyclopedia*. New York: HarperCollins, 2001. ISBN 0-06-273755-4.
- Myrick, Susan, *White Columns in Hollywood: Reports from the GWTW Sets*. Macon, Georgia: Mercer University Press, 1982 ISBN 0-86554-044-6.
- Wakeman, John, *World Film Directors*. New York: H. W. Wilson Company 1987. ISBN 0-8242-0757-2.

External links

- George Cukor (http://www.imdb.com/name/nm2030/) at the Internet Movie Database
- George Cukor (http://www.ibdb.com/person.asp?ID=14534) at the Internet Broadway Database
- George Cukor (http://tcmdb.com/participant/participant.jsp?participantId=41936) at the TCM Movie Database
- Cukor bibliography at UC Berkeley Media Resources Center (http://www.lib.berkeley.edu/MRC/cukor.html)
- Senses of Cinema: Great Directors Critical Database (http://archive.sensesofcinema.com/contents/directors/04/cukor.html)
- Literature on George Cukor (http://www.virtual-history.com/movie/person/2354/george-cukor)

European_Theatre_of_World_War_II

The **European Theatre of World War II** was a huge area of heavy fighting across Europe from Germany's invasion of Poland on September 1, 1939 until the end of the war with the German unconditional surrender on May 8, 1945 (V-E Day). The Allied forces fought the Axis powers in three sub-theatres: the Eastern Front, the Western Front, and the Mediterranean Theatre.

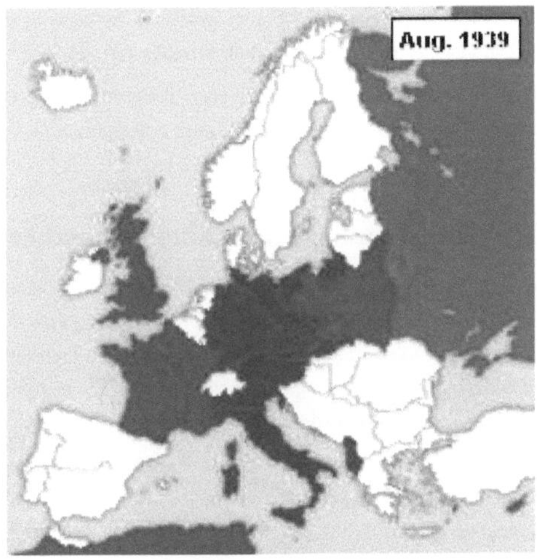

Animation of the WWII European Theatre. (Maps of the front at two-week intervals July 1943 to August 1945 available here)

Preceding events

Germany was defeated in World War I, and the Treaty of Versailles placed punitive conditions on the country, including significant financial reparations, the loss of territory (some only temporarily), war guilt, military weakening and limitation, and economic weakening. Germany was humiliated in front of the world and had to pay very large war reparations. Many Germans blamed their country's post-war economic collapse and hyperinflation on the treaty's conditions. These resentments contributed to the political instability which made it possible for Adolf Hitler and his National Socialist Party to come to power, with Hitler's appointment as Chancellor of Germany in 1933.

After Hitler took Germany out of the League of Nations, Mussolini and Hitler formed the Rome-Berlin axis, under a treaty known as the Pact of Steel. Later, Japan would also join as an Axis power under the rule of Hideki Tojo. Japan and Germany had already signed the Anti-Comintern Pact in 1939, to counter the Soviet Union's communism "threat". Other smaller powers also later joined the Axis through pacts.

Outbreak of war in Europe

Germany and the Soviet Union were sworn enemies, but following the Munich Agreement, which effectively handed over Czechoslovakia (a French and Soviet ally, and the only remaining presidential democracy in Central Europe) to Germany, political realities allowed the Soviet Union to sign a non-aggression pact (the Molotov-Ribbentrop Pact) including a secret clause partitioning Poland, the Baltic Republics and Finland between the two spheres of interests.

Adolf Hitler and Benito Mussolini in Italy

Full-scale war in Europe began at dawn on September 1, 1939, when Germany used her newly formed Blitzkrieg tactics and military strength to invade Poland, to which both the United Kingdom and France had pledged protection and independence guarantees. On September 3, 1939, the UK and France declared war on Germany and British troops were sent to France, however neither French nor British troops gave any significant assistance to the Poles during the entire invasion, and the German-French border, excepting the Saar Offensive, remained mostly calm.

On September 17, the Soviet forces attacked. The Polish government evacuated the country for Romania. Poland fell within five weeks, with her last large operational units surrendering on October 5 after the Battle of Kock. As the Polish September Campaign ended, Hitler offered to Britain and France peace on the basis of recognition of German European continental dominance. On October 12, the United Kingdom formally refused.

Despite the quick campaign in the east, along the Franco-German frontier the war settled into a quiet period. This relatively non-confrontational and mostly non-fighting period between the major powers lasted until May 10, 1940, and was known as the Phoney War.

Germany assumes dominance in northern Europe

Several other countries, however, were drawn into the conflict at this time. By September 28, 1939, the three Baltic Republics felt they had no choice but to permit Soviet bases and troops on their territory. The Baltic Republics were occupied by the Soviet army in June 1940, and finally annexed to the Soviet Union in August 1940.

Finnish soldiers during the Winter War

The Soviet Union wanted to annex Finland and offered a union agreement, but Finland rejected it, which caused Soviet Union to attack Finland on November 30. This began the Winter War. After over three months of hard fighting, and heavy losses, the Soviet Union gave up the attempted invasion. In the Moscow Peace Treaty of March 12, 1940, Finland ceded 10% of her territory. The Finns were embittered over having lost more land in the peace than on the battlefields, and over the perceived lack of world sympathy.

Meanwhile in western Scandinavia, Germany invaded Denmark and Norway in April 1940, and in response, Britain occupied the Faroe Islands (a Danish territory) and invaded and occupied Iceland (a sovereign nation with the King of Denmark as its monarch).

Sweden during World War II was able to remain neutral.

War comes to the west

German troops in Paris after the fall of France.

On May 10 the Phoney War ended with a sweeping German invasion of Belgium, the Netherlands, Luxembourg, and French Third Republic that bypassed French fortifications along the Maginot Line. After overrunning Holland, Belgium, and Luxembourg, Germany turned against France, entering the country through the Ardennes on May 13—the French had made the fatal mistake of leaving this area almost totally undefended, believing its terrain to be impassable for tanks and other vehicles. Most Allied forces were in Flanders, anticipating a re-run of the World War I Schlieffen Plan, and were cut off from the French mainland. As a result of this, and also the superior German communications and tactics, the Battle of France was shorter than virtually all prewar Allied thought could have conceived. It lasted six weeks, including the Luftwaffe bombing of Paris June 3, after which France surrendered on June 22. In order to further the humiliation of the French people and the country itself, Hitler arranged for the surrender document to be signed in the same railway coach where the German surrender had been signed in 1918. The surrender divided France into two parts; the Northern part under French control, and a southern part under German control, based at Vichy and referred to as Vichy France, a rump state allied and friendly to Germany. Many French soldiers, as well as those of other occupied countries, escaped to Britain. The General de Gaulle proclaimed himself the legitimate leader of the Free French organization and vowed to continue to fight. On June 10 Italy declared war on both France and the United Kingdom.

Vyacheslav Molotov, the Foreign Policy Minister of the U.S.S.R., which was tied with Soviet-German non-aggression treaty, congratulated the Germans: *"We hand over the most cordial congratulations by the Soviet government on the occasion of splendid success of German Wehrmacht. Guderian's tanks broke through to the sea near Abbeville, powered by Soviet fuel, the German bombs, that razed Rotterdam to the ground, were filled with Soviet pyroxylin, and bullet cases, which hit the British soldiers retreating from Dunkirk, were cast of Soviet cupronickel alloy..."* [1]

Later, on April 24, 1941, the U.S.S.R. gave full diplomatic recognition to the Vichy government situated in the non-occupied zone in France.[1]

Thus, the fall of France left Britain and its Empire to stand alone. The British Prime Minister, Neville Chamberlain, resigned during the battle and was replaced by Winston Churchill. Fortunately for Britain, much of its army escaped capture from the northern French port of Dunkirk, where hundreds (if not thousands) of tiny civilian boats were used to ferry troops from the beaches to the waiting warships. There is much debate over whether German Panzer divisions could have defeated these soldiers alone if they had pressed forward, since the tank divisions were overextended and would require extensive refitting; in any case, Hitler elected to follow the advice of the leader of German air forces Hermann Göring and allow the Luftwaffe alone to attack the Allied forces until German infantry was able to advance, giving the British a window for the evacuation. Later, many of the evacuated troops would form an important part and the center of the army that landed at Normandy on D-Day.

The British rejected several covert German attempts to negotiate a peace. Germany massed their air force in northern German-occupied France to prepare the way for a possible invasion, codenamed *Operation Seelöwe* (Sea Lion), deeming that air superiority was essential for the invasion. The operations of the Luftwaffe against the Royal Air Force became known as the Battle of Britain. Initially the Luftwaffe concentrated on destroying the R.A.F. on the ground and in the air. They later switched to bombing major and large industrial British cities in the Blitz, in an attempt to draw R.A.F. fighters out and defeat them completely. Neither approach was successful in reducing the R.A.F. to the point where air superiority could be obtained, and plans for an invasion were suspended by September 1940.

During the Blitz, all of Britain's major industrial, cathderal, and political cites were heavily bombed. London suffered particularly, being bombed each night for several months. Other targets included Birmingham and Coventry, and strategically important cities, such as the naval base at Plymouth and the port of Kingston upon Hull. With no land forces in direct conflict in Europe, the war in the air attracted worldwide attention even as sea units fought the Battle of the Atlantic and a number of British commando raids hit targets in occupied Europe. Churchill famously said of the R.A.F. personnel who fought in the battle: *"Never in the field of human conflict was so much owed by so many to so few".—*

The war in the air

The RAF Supermarine Spitfire, used extensively during the Battle of Britain.

The air war in the European Theatre commenced in 1939.

Prewar doctrine had held that waves of bombers hitting enemy cities would cause mass panic and the rapid collapse of the enemy. As a result, the Royal Air Force had built up a large strategic bomber force. By way of contrast, Nazi German air force doctrine was almost totally dedicated to supporting the army. Therefore, German bombers were smaller than their British equivalents, and Germany never developed a fully successful four engined heavy bomber equivalent to the Lancaster or B-17, with only the similarly sized Heinkel He 177 placed into production and made operational for such duties with the *Luftwaffe* in the later war years.

The main concentration of German raids on British cities was from September 7, 1940 until May 10, 1941 in the most famous air battle of all time, known as the Battle of Britain. Facing odds of four against one the RAF held off the mighty Luftwaffe forcing Hermann Wilhelm Göring to withdraw his forces and more importantly indefinitely postpone invasion plans. This proved the first major turning point of the War. After that most of the strength of the Luftwaffe was diverted to the war against the Soviet Union leaving German cities vulnerable to British and later American air bombings. As a result of the victory, Great Britain was used by U.S and other Allied forces as a base from which to begin the D-Day landings in June 1944 and the liberation of Nazi-occcupied Western Europe. Neththeless German raids continued on British cities albeit on a smaller and less destructive scale for the rest of the war, and later the V1 Flying Bomb and V-2 ballistic missile were both used against Britain. However, the balance of bomb tonnage being dropped shifted greatly in favour of the RAF as Bomber Command gained in strength. By 1942, Bomber Command could put 1,000 bombers over one German city.

From 1942 onwards, the efforts of Bomber Command were supplemented by the Eighth Air Force of the United States Army Air Forces, U.S. Army Air Forces units being deployed to England to join the assault on mainland Europe on July 4, 1942. Bomber Command raided by night and the US forces by day. On February 14, 1945, a raid on Dresden produced one of the most devastating fires in history. A firestorm was created in the city, and between 18,000 to 25,000 people were killed.[2] [3] [4] Only the raids on Hamburg (July 24, 1943 – July 29, 1943), the March 9–10, 1945 firebombing of Tokyo and the nuclear attacks on Hiroshima (August 6, 1945) and Nagasaki (August 9, 1945) killed more people through a single attack.

The Mediterranean and Balkans

Italy had invaded Albania on April 7, 1939 and had officially annexed it. Mussolini's regime declared war on Britain and France on June 10, 1940, and invaded Greece on October 28. However, Italian forces were unable to match the Nazi successes in northwest Europe.

The Italian *Regia Aeronautica* began the long and unsuccessful siege of Malta on June 12. Even the surrender of France did not greatly assist the Axis forces. The naval Battle of the Mediterranean was a disaster for the Italian *Regia Marina* and the Vichy French navy, which were effectively destroyed as fighting forces by the Royal Navy and the Royal Australian Navy during 1940, most notably in the attack on Mers-el-Kebir (July 3) and the Battle of Taranto (November 11).

Not only did the Italians fail to conquer Greece, but under the supervision of Greece's dictator, Prime Minister Ioannis Metaxas, the Greeks successfully counterattacked into Albania, from November 14.

The Kingdom of Yugoslavia was lacking true leadership of a king and instead ruled by a regency headed by Prince Pavle Karađorđević, signed the Tripartite Treaty on March 25, 1941. The regency did this because they were promised by Hitler that if they joined and let the Axis attack Greece through their territory Yugoslavia would be given areas of Northern Greece including Salonika. However, soon afterwards, after public demonstrations, a March 27 coup d'état was made by Army General Dušan Simović which took control away from the regency and distanced Yugoslavia from the fascists.

The imminent Greek victory over Italy prompted German intervention. On April 6, 1941 German forces, supported by the Italians, Hungarians and the Bulgarians, engaged in combat with the Greeks and simultaneously invaded Yugoslavia. British, Australian and New Zealand forces were hastily dispatched from Egypt to Greece, but the Allies lacked a co-ordinated strategy, were comprehensively beaten and evacuated to Crete. Advancing rapidly, Axis forces captured Athens, Greece's capital on April 27, 1941 effectively placing most of the country under occupation. The reinforcement of the Greeks weakened the British forces in Libya such that they were unable to complete the destruction of Italian forces there before they were reinforced by the German *Afrika Korps* prolonging the war in North Africa.

After the mainland was conquered, Germany invaded Crete in what is known as the Battle of Crete (May 20, 1941 – June 1, 1941). Instead of an amphibious assault as expected, the Germans mounted a large airborne invasion. The paratroopers (*Fallschirmjäger*) suffered severe losses and large-scale airborne operations were given up after that. However, the Germans eventually prevailed on Crete. Most of the Allied forces were evacuated to Egypt — joining King George II of Greece and the exiled Greek government of Emmanouil Tsouderos — on June 1, 1941.

Once the Balkans were secure, the largest land operation in history was launched, when Germany attacked the Soviet Union. The Balkans campaign delayed the invasion, and subsequent resistance movements in Albania, Yugoslavia and Greece tied up valuable Axis forces. This provided much needed and possibly decisive relief for the Soviets.

The Eastern Front

On June 22, 1941, Germany launched the invasion of the Soviet Union, code-named Operation Barbarossa.[5] This invasion, the biggest in recorded history, started the bloodiest conflict in world history; the Axis-Soviet War. The Eastern Front was by far the largest and bloodiest theatre of World War II. It is generally accepted as being the most lethal conflict in human history, with over 30 million dead as a result. It involved more land combat than all other World War II theatres combined. The distinctly brutal nature of warfare on the Eastern Front was exemplified by an often willful disregard for human life by both sides.

The leader of the USSR, Joseph Stalin, had been warned repeatedly by outside sources and his own intelligence network of the impending invasion, but he ignored the warnings due to conflicting information presented to him by the Soviet intelligence. Moreover, on the very night of the invasion Soviet troops received a directive undersigned by Marshal Timoshenko and General of the Army Georgi Zhukov that commanded: "do not answer to any provocations" and "do not undertake any actions without specific orders". The early weeks of the invasion were devastating for the Soviet Army. Enormous numbers of Soviet troops were encircled in pockets and fell into Nazi German hands. In addition to German troops, a few Italian, Hungarian and Romanian troops were also involved in the campaign. Finland also sent troops, but oddly, the Finns initially declared neutrality, however with both German and Soviet troops on her soil, Finland was well prepared to join forces with Germany when the Soviet Union attacked on June 25. The following conflict from 1941 to 1944 is sometimes referred to as the Continuation War, as in the continuation of the Winter War.

Operation Barbarossa suffered from several fundamental flaws. The most serious of these was the logistical situation of the attack. The sheer vastness of the distances in the Soviet Union meant that Germany could only advance so far before outrunning their supply chains. By the time the German attack froze to a halt before Moscow on December 5, 1941, it literally could not go any further. There simply were not enough supplies reaching the front to conduct proper defensive operations, let alone a proper offense. The timetable that Barbarossa was planned to assumed that the Soviets would collapse before the Russian winter hit. The failure of that to happen also fatally affected Nazi German plans. Had Hitler not invaded Greece and Yugoslavia earlier in the year, the invasion would have proceeded at that time, and the Soviet Union might have collapsed.

During their long retreat, the Soviets employed a scorched earth policy. They burnt crops and destroyed utilities as they withdrew before Germany. That helped to contribute to the logistical problems that Germany experienced. More importantly for them, the Soviets also succeeded in a massive and unprecedented removal of their industry from the threatened war zone to protected areas in the East.

The extension of the campaign beyond the length that Germany expected meant that the German Army suffered hundreds of thousands of casualties in the bitter cold of the Russian winter, and from the counterattacks of Soviet units.

Even with their advance having ground to a halt due to a lack of supplies and the onset of winter, Germany had conquered a vast amount of territory, including two-fifths of the Soviet economy. Dislodging them proved difficult and eventually cost the Soviet Union dearly.

A few months after the invasion began, German troops laid siege to Leningrad (known as the Siege of Leningrad) from the North by Finnish forces, and from the South by the German Wehrmacht. Finland's C-in-C Mannerheim had halted at the River Svir and refrained from attacking the city. Hitler had ordered that the city of Leningrad must "vanish from the surface of the earth", with its entire population exterminated. Rather than storming the city, the Wehrmacht was ordered to blockade Leningrad so as to starve the city to death, while attacking it with bombers and artillery. About one million civilians died in the Leningrad siege – 800,000 by starvation. The siege lasted for 872 days. During the winter, the only way into the city was across Lake Ladoga, between the German and Finnish lines.

After enduring the winter of 1941–1942, the German army prepared for further offensive operations. One of the major problems faced by the Nazi war machine in World War II was a shortage of oil. For this reason, Germany decided to give up on Moscow for the time being, and the summer offensive of 1942 decided to focus on the war in the south, with the target being the oil fields of the Caucasus. In a major blunder, Hitler split Army Group South into two subgroups, Army Group A which would attack the Caucasus and Army Group B which would advance towards the city of Stalingrad (now Volgograd).

The "Big Three" Allied leaders. From left to right: Winston Churchill (UK), Franklin Roosevelt (USA), and Joseph Stalin (USSR).

Indecision by Hitler, dissent among the higher-ranked Nazi German officers, and extended supply lines combined in a prolonged battle in the streets of Stalingrad. Germany eventually occupied over 90% of the city, but in an attempt to defeat the remaining Soviet defenders almost all German soldiers in the area were funnelled into the ruins of the city. Months of bitter hand-to-hand combat in the ruins of the city depleted the German forces, leaving only weak Romanian and Hungarian forces to guard the flanks of the Stalingrad army group. In Operation Uranus, the Soviets easily defeated these minor Axis forces as they performed a massive encirclement operation. The German troops remaining in the city were trapped – cut off from their supply lines and starving, they were ordered by Hitler to fight to the last man, and they displayed incredible fortitude and bravery under unbearable conditions.

Starved of food, fuel, ammunition, and clothes, the pocket was gradually reduced, with the last portion surrendering on February 2, 1943. In a cynical attempt to prevent the surrender, Hitler promoted Friedrich Paulus, commander of the 6th Army to Field Marshal, because no German of that rank had ever surrendered. Heavy losses affected both sides in the Battle of Stalingrad, one of the bloodiest battles in history. About 1.5 million people perished in this battle, including 100,000 civilians in the city.

After Stalingrad, the initiative had passed from Germany but had not yet been seized by the Soviets. A desperate counterattack in the spring of 1943 by forces of Field Marshal Erich von Manstein temporarily halted the Soviet advance, and led to the largest tank battle in history, at Kursk. Kursk was the last major offensive by the German Army on the eastern front. The Soviets had intelligence of what was to come and prepared massive defences in huge depth in the Kursk salient. They stopped the German armoured thrusts after a maximum penetration of just over 30 miles (km). After Kursk the Red Army never ceased being on the offensive until Berlin was captured in May 1945.

More Soviet citizens died during World War II than those of all other countries combined. Nazi ideology considered Slavs to be "subhuman" and German forces committed ethnically targeted mass murder. Civilians were rounded up and burned alive or shot in squads in many cities conquered by the Nazis. Approximately 27 million Soviets, among them more than 20 million civilians in Soviet cities and areas, were killed in the Nazi invasion of the Soviet Union.

At least 7 million Red Army troops died facing the Germans and their allies in the Eastern Front. The Axis forces themselves had lost over 6 million troops, whether by combat or by wounds, disease, starvation or exposure; another several hundred thousand were seized as POWs and over half died in Soviet gulags because of disease, starvation, or shortage of supplies.

Lend-Lease supplies from the United Kingdom and the United States made very important impact for Soviet military forces. Supply convoys sailed to Soviet ports that were patrolled by Nazi U-boats. Allied activities before D-Day may have tied up only a few divisions in actual fighting, but many more were forced to guard lonely coasts against raids that never came or to man antiaircraft guns throughout Europe.

Allied invasion of Italy

Successes in the North African desert left the Allies in complete control of the Mediterranean's southern shore and allowed the Allied Forces Headquarters to start planning an attack into what Winston Churchill referred to as the "soft underbelly" of Europe.

The Allies first action was the capture of the island of Sicily, called Operation Husky, on July 10, 1943. This brought to the fore a growing dissatisfaction with Mussolini. On July 29, 1943, the King of Italy fired Mussolini, and placed him under house arrest in an isolated mountain resort. His replacement, General Pietro Badoglio, negotiated an armistice with the Allies on September 8, 1943. Germany moved quickly into the confused situation, disarming Italian formations, taking military control of Italian areas, and preparing to defend Italy on her own.

Allied troops landed on mainland southwestern Italy on September 3, 1943, crossing from Sicily. Further landings were made at Salerno and Taranto on September 9. For more information see: Allied invasion of Italy. This led Italy, already angry at its former leader Mussolini, to join the Western Allies.

A German commando raid Operation Eiche, led by Otto Skorzeny, rescued the imprisoned Mussolini. The Germans installed him as the "Head of State" and "Minister of Foreign Affairs" of a Nazi puppet state in northern Italy called the Italian Social Republic.

Germany had built a number of defensive lines through the mountains; the main one was called the Winter Line. The Allies came up against this in the winter of 1943 and were unable to break through. Amphibious landings at Anzio were made in an attempt to bypass the line: however the landing forces were contained by the Germans, and the Gustav Line (the core part of the Winter Line defences) remained intact. Finally, the line was broken in May 1944, in the fourth major attempt in four months to open the road to Rome dominated by the strategically positioned and historic Benedictine Abbey at Monte Cassino.

The Allies finally liberated Rome on June 4, 1944, two days before the landings in Normandy. Germany made a fighting withdrawal to the Gothic Line north of Florence. From September 10, 1944 till the end of the year Allied forces attacked the line and in some of the fiercest fighting of the war, broke the Gothic defences but failed to break through to the Lombardy Plain. The offensive by Allied and some Italian forces resumed when the weather permitted in early April 1945 and continued until Germany surrendered in Italy on April 29, 1945 two days after Mussolini's Allied capture and one day before Hitler's suicide.

Allied invasion of occupied France

Simultaneously with the fall of Rome came the long-awaited invasion of France. Operation Overlord put over 180,000 troops ashore in Normandy on June 6, 1944, creating a beachhead that would eventually result in over 3 million Allied soldiers on Germany's western front. A long grinding campaign six weeks long followed as American, British, and Canadian forces were slowly built up in the beachhead, and German forces slowly worn down. When the breakout finally did come it was spectacular, with Allied troops very quickly capturing almost all of Normandy within days. Many German forces that had been fighting in Normandy were trapped in the Falaise pocket.

General Charles de Gaulle, leader of the Free French in opposition to Pétain's Vichy regime.

Incessant bombing of Germany's infrastructure and cities caused tremendous casualties and disruption. Internally, Hitler survived a number of Nazi inner assassination attempts. The most serious was the July 20 Plot, occurring on July 20, 1944. Orchestrated by Claus von Stauffenberg and involving among others Erwin Rommel and Alfred Delp, the plot had intended to place a time bomb in a position to kill Hitler but a number of unscheduled factors and operation failures led to its failure. Hitler was only slightly injured.

Operation Overlord was complemented by an invasion of southern France on August 15, 1944 codenamed Operation Dragoon. By September 1944 three Allied Army Groups were in line against German formations in the west. There was optimism that the war in Europe might be over by the end of 1944.

An attempt was made to force the situation with Operation Market Garden (September 17, 1944 – September 25, 1944). The Allies attempted to capture bridges with an airborne assault, to open the way into Germany and liberate the northern Netherlands. Since heavier German forces than intelligence had predicted were present, the British 1st Airborne Division was almost completely destroyed, and the operation failed.

The weather of 1944 combined with a poor situation for the Allies led to a stagnant situation on the western front. The Americans continued to grind away at the defenders in the Battle of Hurtgen Forest (September 19, 1944 – February 10, 1945). As long as Germany stayed on the defence, the Allies were hard-pressed to advance rapidly.

That changed when Germany mounted a major counteroffensive on December 16, 1944. The Ardennes offensive, also called the Battle of the Bulge, drove back and surrounded some small American units. The Allied forces were eventually successful in driving back Germany, in what turned out to be their last major advance of the war. The battle officially ended on January 27, 1945.

The final obstacle to the Allies was the Rhine. It was crossed in March 1945, and the way lay open to the center of Germany. The last major German forces in the west were encircled and trapped in the Ruhr.

End of the war in Europe

see also Victory in Europe Day, Victory Day (Eastern Front), Western Front (World War II).

Prime Minister Winston Churchill waves to crowds in London on Victory in Europe Day.

On April 27, 1945, as Allied forces closed in on Milan, Mussolini was captured by Italian Partisans. He was trying to flee Italy to Switzerland and was traveling with a German anti-air battalion. On April 28, Mussolini and several of the other Fascists captured with him were taken to Dongo and executed by squad shooting. The bodies were then taken to Milan and unceremoniously strung up in front of a filling station.

Hitler, learning of Mussolini's death, realized that the end had finally come. He remained in Berlin, the last Nazi held city and the crumbling Nazi Empire's capitol, even as the city was encircled and trapped by the Soviets and the Battle of Berlin raged. On April 30, Adolf Hitler, with his wife of one day, Eva Braun, committed suicide in his bunker to avoid capture by Soviet troops. In his last will and testament, Hitler appointed Grand Admiral Karl Dönitz as the new German leader. But Germany lasted only 7 days longer under the "Flensburg government" of Dönitz. He surrendered unconditionally to the Americans, British, and Soviets on May 8, 1945 ("VE Day").

In late July and August 1945 the Potsdam Conference finally disbanded, denazified and demilitarized the former Nazi German state and reversed all German annexations and occupied territories.

See also

Atlas of the World Battle Fronts		
 1943-07-01	 1943-11-01	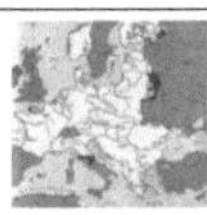 1944-07-01
 1944-09-01	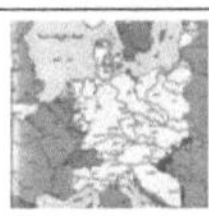 1944-12-01	 1945-05-01

- European-African-Middle Eastern Campaign Medal
- Mediterranean Theatre of Operations
- Pacific War

Citations and notes

[1] The Isolation of the Revolution (http://home.pacific.net.au/~andy/bs/bs1-2.htm)
[2] Bojan Pancevski Dresden bombing death toll lower than thought (http://www.telegraph.co.uk/news/newstopics/onthefrontline/3123512/Dresden-bombing-death-toll-lower-than-thought.html), Daily Telegraph, 3 October 2008.
[3] Kate Connolly, Panel rethinks death toll from Dresden raids (http://www.guardian.co.uk/world/2008/oct/03/secondworldwar.germany), The Guardian, 3 October 2008.
[4] Landeshauptstadt Dresden (1 October 2008). "Erklärung der Dresdner Historikerkommission zur Ermittlung der Opferzahlen der Luftangriffe auf die Stadt Dresden am 13./14. Februar 1945" (http://www.dresden.de/media/pdf/presseamt/Erklaerung_Historikerkommission.pdf). Landeshauptstadt Dresden. . Retrieved 13 February 2010.
[5] Amnon Sella. 'Barbarossa': Surprise Attack and Communication. Journal of Contemporary History, Vol. 13, No. 3, (Jul., 1978), pp. 555–583.

Further reading

- Churchill, Winston (1948–1953), *The Second World War*, 6 vols.
- Keegan, John (1989). *The Second World War*, Hutchinson. ISBN 0-09-174011-8.
- Lee, Loyd E. ed. *World War II in Europe, Africa, and the Americas, with General Sources: A Handbook of Literature and Research* (1997) (http://www.questia.com/PM.qst?a=o&d=27558710)
- Murray, Williamson and Millett, Allan R. (2000). *A War to Be Won: Fighting the Second World War*, Harvard University Press. ISBN 0-674-00163-X.
- Overy, Richard (1995). *Why the Allies Won* Pimlico. ISBN 0-7126-7453-5.
- Powaski, Ronald E. (2003). *Lightning War: Blitzkrieg in the West, 1940*. John Wiley. ISBN 0-471-39431-9, 9780471394310.
- Powaski, Ronald E. (2008). *Lightning War: Blitzkrieg in the West, 1940*. Book Sales, Inc.. ISBN 0-7858-2097-3, 9780785820970.
- Smith, J. Douglas and Richard Jensen. (2002) *World War II on the Web: A Guide to the Very Best Sites*
- Weinberg, Gerhard L. (2005). *A World at Arms: A Global History of World War II,* Cambridge University Press. ISBN 0-521-44317-2.

Mickey_Rooney

Mickey Rooney	
 Rooney in 1945	
Born	Joseph Yule, Jr.September 23, 1920Brooklyn, New York, U.S.
Residence	Westlake Village, California
Occupation	Actor, comedian
Years active	1922–present
Height	5'2" (1.57 m)
Spouse(s)	Ava Gardner (1942–43; divorced) B.J. Baker (1944–48; divorced) Martha Vickers (1949–51; divorced) Elaine Devry (1952–58; divorced) Carolyn Mitchell (1958–66; her death) Marge Lane (1966–67; divorced) Carolyn Hockett (1969–75; divorced) Jan Chamberlin (1978–present)
Children	9
Parents	Joseph Yule, Nellie W. (née Carter)
Awards	Academy Juvenile Award, Academy Honorary Award, Academy Award, 2 Golden Globe Awards, Emmy Award
Website	
mickeyrooney.com [1]	

Mickey Rooney (born **Joseph Yule, Jr.**; September 23, 1920) is an American film actor and entertainer whose film, television, and stage appearances span nearly his entire lifetime.

He has received multiple awards, including a Juvenile Academy Award, an Honorary Academy Award, two Golden Globes and an Emmy Award. Working as a performer since he was a child, he was a superstar as a teenager for the films in which he played Andy Hardy, and he has had one of the longest careers of any actor, to date spanning 90 years actively making films in ten decades, from 1920s to 2010s. For a younger generation of fans, he gained international fame for his leading role as Henry Dailey in The Family Channel's *The Adventures of the Black Stallion*, as well as the film itself.

Along with Carla Laemmle, Lupita Tovar, and Baby Peggy, he is one of the last surviving stars who worked in the silent film era.

Early life

Rooney was born Joseph Yule, Jr. in Brooklyn, New York. His father, Joseph Yule, was from Glasgow, Scotland, and his mother, Nellie W. (née Carter), was from Kansas City, Missouri. Both of his parents were in vaudeville, appearing in a Brooklyn production of *A Gaiety Girl* when Joseph, Jr. was born. He began performing at the age of 17 months as part of his parents' routine, wearing a specially tailored tuxedo.[1]

When he was 14 months old, unknown to everyone, he crawled onstage wearing overalls and a little harmonica around his neck. He sneezed and his father, Joe Sr., grabbed him up, introducing him to the audience as Sonny Yule. He felt the spotlight on him and has described it as his mother's womb. From that moment on, the stage was his home.

His father was a womanizer and a heavy drinker, leaving the family when Joe Jr. was only three. While Joe Sr. was traveling, Joe Jr. and his mother moved from Brooklyn, New York to Kansas City, Missouri to live with his aunt. While his mother was reading the entertainment newspaper, Nellie was interested in getting Hal Roach to approach the young star to participate in the *Our Gang* series in Hollywood. Roach offered $5 a day to Joe Jr. while the other young stars were paid five times more.

As he was getting bit parts in films, he was working with other established film stars such as Joel McCrea, Douglas Fairbanks, Jr. and Jean Harlow. While selling newspapers around the corner, he also entered into Hollywood Professional School, where he went to school with dozens of unfamiliar students such as: Joseph A. Wapner, Nanette Fabray, Judy Garland, Lana Turner, among many others, and later Hollywood High School, where he graduated in 1938.

Career

Mickey McGuire

The Yules separated in 1924 during a slump in vaudeville, and in 1925, Nell Yule moved with her son to Hollywood, California, where she managed a tourist home. Fontaine Fox had placed a newspaper ad for a dark-haired child to play the role of "Mickey McGuire" in a series of short films. Lacking the money to have her son's hair dyed, Mrs. Yule took her son to the audition after applying burnt cork to his scalp.[2] Joe got the role and became "Mickey" for 78 of the comedies, running from 1927 to 1936, starting with *Mickey's Circus*, released September 4, 1927.[3] These had been adapted from the *Toonerville Trolley* comic strip, which contained a character named Mickey McGuire. Joe Yule briefly became Mickey McGuire legally in order to trump an attempted copyright lawsuit (if it were his legal name, the film producer Larry Darmour did not owe the comic strip writers royalties). His mother also changed her surname to McGuire in an attempt to bolster the argument, but the film producers lost. The litigation settlement awarded damages to the owners of the cartoon character, as well as compelled the twelve-year-old actor to refrain from calling himself by the name Mickey McGuire on and off screen.[4]

Rooney later claimed that, during his Mickey McGuire days, he met cartoonist Walt Disney at the Warner Brothers studio, and that Disney was inspired to name Mickey Mouse after him,[5] although Disney always said that he had changed the name from "Mortimer Mouse" to "Mickey Mouse" on the suggestion of his wife.[6]

During an interruption in the series in 1932, Mrs. Yule made plans to take her son on a ten-week vaudeville tour as McGuire, and Fox sued successfully to stop him from using the name. Mrs. Yule suggested the stage name of Mickey Looney for her comedian son, which he altered slightly to Rooney, a less frivolous version.[2] Rooney did other films in his adolescence, including several more of the McGuire films, and signed with Metro-Goldwyn-Mayer in 1934. MGM cast Rooney as the teenage son of a judge in 1937's *A Family Affair*, setting Rooney on the way to another successful film series.

"Andy Hardy" and Judy Garland

Rooney with Judy Garland in *Love Finds Andy Hardy* (1938)

In 1937, Rooney was selected to portray Andy Hardy in *A Family Affair* (1937), which MGM had planned as a B-movie.[2] Rooney provided comic relief as the son of Judge James K. Hardy, portrayed by Lionel Barrymore (although Lewis Stone would play the role of Judge Hardy in subsequent films). The film was an unexpected success, and led to 13 more *Andy Hardy* films between 1937 and 1946, and a final film in 1958. Rooney also received top billing as "Shockey Carter" in *Hoosier Schoolboy* (1937).

Also in 1937, Rooney made his first film alongside Judy Garland with *Thoroughbreds Don't Cry*. Garland and Rooney became close friends and a successful song and dance team. Besides three of the Andy Hardy films, where she portrayed Betsy Booth, a younger girl with a crush on Andy, they appeared together in a string of successful musicals, including the Oscar-nominated *Babes in Arms* (1939). During an interview in the documentary film *When the Lion Roars*, Rooney describes their friendship:

> Judy and I were so close we could've come from the same womb. We weren't like brothers or sisters but there was no love affair there; there was more than a love affair. It's very, very difficult to explain the depths of our love for each other. It was so special. It was a forever love. Judy, as we speak, has not passed away. She's always with me in every heartbeat of my body.[7]

Rooney's breakthrough role as a dramatic actor came in 1938's *Boys Town* opposite Spencer Tracy as Whitey Marsh, which opened shortly before his 18th birthday. Rooney was awarded a special Juvenile Academy Award in 1939[8] and was named the biggest box-office draw in 1939, 1940 and 1941.[9] Unquestionably a well-known entertainer by the early 1940s, his picture appeared on the cover of the March 18, 1940 issue of Time Magazine, one of the few show business personalities so honored in that era. Rooney, with Garland, was one of many celebrities caricatured in Tex Avery's 1941 Warner Bros. cartoon *Hollywood Steps Out*. As of 2013, Rooney is the only surviving entertainer depicted in the cartoon. In 1991, Rooney was honored by the Young Artist Foundation with its Former Child Star "Lifetime Achievement" Award recognizing his achievements within the film industry as a child actor.[10] After presenting the award to Rooney, the foundation subsequently renamed the accolade "The Mickey Rooney Award" in his honor.[11] [12]

After the war

In 1944, Rooney entered military service. He served more than 21 months, until shortly after the end of World War II. During and after the war he helped entertain the troops in America and Europe, and spent part of the time as a radio personality on the American Forces Network and was awarded the Bronze Star Medal for entertaining troops in combat zones. In addition to the Bronze Star Medal, Rooney also received the Army Good Conduct Medal, American Campaign Medal, European-African-Middle Eastern Campaign Medal, and World War Two Victory Medal for his military service.

Rooney entertaining troops in 1945

After his return to civilian life, his career slumped. He appeared in a number of films, including *Words and Music* in 1948, which paired him for the last time with Garland on film (he appeared with her on one episode as a guest on her CBS variety series in 1963). He briefly starred in a CBS radio series, *Shorty Bell*, in the summer of 1948, and reprised his role as "Andy Hardy", with most of the original cast, in a syndicated radio version of *The Hardy Family* in 1949 and 1950 (repeated on Mutual during 1952).[13]

His first television series, *The Mickey Rooney Show: Hey, Mulligan* (created by Blake Edwards with Rooney as his own producer), appeared on NBC television for thirty-two episodes between August 28, 1954 and June 4, 1955. In 1951, he directed a feature film for Columbia Pictures, *My True Story* starring Helen Walker. Rooney also starred as a ragingly egomaniacal television comedian in the live 90-minute television drama *The Comedian*, in the *Playhouse 90* series on the evening of Valentine's Day in 1957, and as himself in a revue called *The Musical Revue of 1959* based on the 1929 film *The Hollywood Revue of 1929* which was edited into a film in 1960, by British International Pictures.

In 1958, Rooney joined Dean Martin and Frank Sinatra in hosting an episode of NBC's short-lived *Club Oasis* comedy and variety show. In 1960, Rooney directed and starred in *The Private Lives of Adam and Eve*, an ambitious comedy known for its multiple flashbacks and many cameos. In the 1960s, Rooney returned to theatrical entertainment. He still accepted film roles in undistinguished films, but occasionally would appear in better works, such as *Requiem for a Heavyweight* (1962), *It's A Mad, Mad, Mad, Mad World* (1963), and *The Black Stallion* (1979). One of Rooney's more controversial roles came in the highly acclaimed 1961 film *Breakfast at Tiffany's* where he played a stereotyped buck-toothed myopic Japanese neighbor (Mr. Yunioshi) of the main character, Holly Golightly. Despite Rooney's protests that he was congratulated for the role by Asians, that role would later be held up as one of the most notorious examples of Hollywood's history of stereotypical depictions of that racial group.

On December 31, 1961, he appeared on television's *What's My Line* and mentioned that he had already started enrolling students in the MRSE (Mickey Rooney School of Entertainment). His school venture never came to fruition. This was a period of professional distress for Rooney; as a childhood friend, director Richard Quine put it: "Let's face it. It wasn't all that easy to find roles for a 5-foot-3 man who'd passed the age of Andy Hardy."[14] In 1962, his debts had forced him into filing for bankruptcy.[15]

In 1966, while Rooney was working on the film *Ambush Bay* in the Philippines, his wife Barbara Ann Thomason (akas: Tara Thomas, Carolyn Mitchell), a former pin-up model and aspiring actress who had won 17 straight beauty contests in Southern California, was found dead in their bed. Beside her was her lover, Milos Milos, an actor friend of Rooney's. Detectives ruled it murder-suicide, which was committed with Rooney's own gun.[16]

Rooney was awarded an Academy Juvenile Award in 1938, and in 1983 the Academy of Motion Picture Arts and Sciences voted him their Academy Honorary Award for his lifetime of achievement. He was mentioned in the 1972 song "Celluloid Heroes" by The Kinks: *"If you stomped on Mickey Rooney/ He'd still turn round and smile..."*

Character actor

In addition to his movie roles, Rooney made numerous guest-starring roles as a character actor for nearly six decades, beginning with an episode of *Celanese Theatre*. The part led to other roles on such television series as *Schlitz Playhouse*, *Playhouse 90*, *Producers' Showcase*, *Alcoa Theatre*, *Wagon Train*, *G.E. True Theater*, *Hennesey*, *The Dick Powell Theatre*, *Arrest and Trial*, *Burke's Law*, *Combat!*, *The Fugitive*, *Bob Hope Presents the Chrysler Theatre*, *The Jean Arthur Show*, *The Name of the Game*, *Dan August*, *Night Gallery*, *The Love Boat*, *Kung Fu: The Legend Continues*, among many others.

Rooney on *The Red Skelton Show*, 1962

Television, stage and *The Black Stallion*

Rooney made a successful transition to television and stage work. In 1961, he guest starred in the 13-week James Franciscus adventure–drama television series *The Investigators* on CBS. In 1963, he even entered *The Twilight Zone*, giving a one-man performance in the episode "The Last Night of a Jockey". In 1964, he launched another half-hour sitcom, *Mickey*, on ABC. The story line had "Mickey" operating a resort hotel in southern California. Son Tim Rooney appeared as Rooney's teenaged son on the program, and Emmaline Henry starred as Rooney's wife. It lasted 17 episodes, ending primarily due to the suicide of co-star Sammee Tong in October 1964.[17]

He won a Golden Globe and an Emmy Award for his role in 1981's *Bill*. Playing opposite Dennis Quaid, Rooney's character was a mentally challenged man attempting to live on his own after leaving an institution. He reprised his role in 1983's *Bill: On His Own*, earning an Emmy nomination for the role.[3]

Rooney did the voices for four Christmas TV animated/stop action specials: *Santa Claus Is Comin' to Town* (1970), *The Year Without a Santa Claus* (1974), *Rudolph and Frosty's Christmas in July* (1979), and *A Miser Brothers' Christmas* (2008)—always playing Santa Claus.

He continued to work on stage and television through the 1980s and 1990s, appearing in the acclaimed stage play *Sugar Babies* with Ann Miller beginning in 1979. Following this, he toured as Pseudelous in Stephen Sondheim's *A Funny Thing Happened on the Way to the Forum*. In the 1990s, he returned to Broadway for the final months of *Will Rogers Follies*, playing the ghost of Will's father. On television, starred in the short-lived sitcom, *One of the Boys*, along with 2 unfamiliar young stars, Dana Carvey and Nathan Lane, in 1982. He toured Canada in a dinner theatre production of *The Mind with the Naughty Man* in the mid-1990s. He played The Wizard in a stage production of *The Wonderful Wizard of Oz* with Eartha Kitt at Madison Square Garden. Kitt was later replaced by Jo Anne Worley. In 1995 he starred with Charlton Heston, Peter Graves and Deborah Winters in the Warren Chaney docudrama *America: A Call to Greatness*.[18] He also appeared in the documentaries *That's Entertainment!* and *That's Entertainment! III*, in both films introducing segments paying tribute to Judy Garland.

Rooney voiced Mr. Cherrywood in *The Care Bears Movie* (1985), and starred as the Movie Mason in a Disney Channel Original Movie family film 2000's *Phantom of the Megaplex*. He had a guest spot on an episode of *The Golden Girls* as Sophia's boyfriend "Rocko", who claimed to be a bank robber. He voiced himself in the *Simpsons* episode "Radioactive Man" of 1995. In 1996–97, Rooney played Talbut on the TV series, *Kleo The Misfit Unicorn*. He co-starred in *Night at the Museum* in 2006 with Dick Van Dyke and Ben Stiller; Rooney filmed a cameo with Van Dyke for the 2009 sequel, *Night at the Museum: Battle of the Smithsonian* which was cut from the film but included as an extra on the DVD release.

Actor Mickey Rooney speaks at the Pentagon in 2000 during a ceremony honoring the USO

After starring in one unsuccessful TV series and for turning down an offer on a huge TV series, Rooney finally hit the jackpot, at 70, when he was offered a starring role on The Family Channel's *The Adventures of the Black Stallion*, where he reprised his role as Henry Dailey from the film of the same name, eleven years earlier. The show was based on a novel by Walter Farley. For this role, he had to travel to Vancouver. Just like the film itself, the *Black Stallion* TV series, Rooney became one of the most beloved stars, that the show itself became an immediate hit with teenagers, young adults and people all over the world. The show was also seen in 70 countries.

In December 2009, he appeared as a guest to a dinner party hosted by David Gest on *Come Dine With Me*.[19]

Current work

Rooney appeared in television commercials for Garden State Life Insurance Company in 1999, alongside his wife Jan Rooney. In commercials shown in 2007, he can be seen in the background washing imaginary dishes.

In 2003, Rooney and his wife began their association with Rainbow Puppet Productions, providing their voices to the 100th Anniversary production of *Toyland!*, an adaptation of Victor Herbert's *Babes in Toyland*. He created the voice for the Master Toymaker while Jan provided the voice for Mother Goose. Since that time, they have created voices for additional Rainbow Puppet Productions including *Pirate Party* which also features vocal performances by Carol Channing. Both productions continue to tour theaters across the country.

He continues to work in film and tours with his wife in a multi-media live stage production called *Let's Put On a Show!* His first performance of this show after the September 11, 2001 terrorist attack was in Bend, Oregon, in which Mickey and Jan requested the show begin with the singing of the "Star Spangled Banner" by Jan offstage with only the American flag visible on stage.

On May 26, 2007, he was grand marshal at the Garden Grove Strawberry Festival. Rooney made his British pantomime debut, playing Baron Hardup in Cinderella, at the Sunderland Empire Theatre over the 2007 Christmas period,[20] [21] a role he reprised in 2009 at the Milton Keynes theatre.[22]

In 2008, Rooney starred as Chief, a wise old ranch owner, in the independent family feature film *Lost Stallions: The Journey Home*, marking a return to starring in equestrian-themed productions for the first time since the 1990s TV show *Adventures of the Black Stallion*. Even though they've acted together before, *Lost Stallions: The Journey Home* is the sole film to date in which he and Jan portrayed a married couple on screen.

Rooney made a brief cameo appearance in *The Muppets* (2011), making his career span ten decades.

Personal life

Rooney has been married eight times. In the 1950s and 1960s, he was often the subject of comedians' jokes for his alleged inability to stay married. He is currently married to Jan Chamberlin. He has a total of nine children, as well as nineteen grandchildren[23] and several great-grandchildren.

Mickey Rooney in 1986

In 1942, he married future Hollywood starlet Ava Gardner, but the two were divorced well before she became a star in her own right. While stationed in the military in Alabama in 1944, Rooney met and married local beauty queen Betty Jane Phillips. This marriage ended in divorce after he returned from Europe at the end of World War II. His subsequent marriages to Martha Vickers (1949) and Elaine Mahnken (1952) were also short-lived and ended in divorce. In 1958, Rooney married Barbara Ann Thomason, but tragedy struck when she was murdered in 1966. Falling into deep depression, he married Barbara's friend, Marge Lane, who helped him take care of his young children. The marriage lasted only 100 days. He was married to Carolyn Hockett from 1969 to 1974, but financial instability ended the relationship. Finally, in 1978, Rooney married Jan Chamberlin, his 8th wife. As of 2012, they live in Westlake Village, California.

Rooney is an outspoken advocate for veterans and senior rights.

On September 23, 2010, Rooney celebrated his 90th birthday at Feinstein's at Loews Regency in the Upper East Side of New York City. Among the people who were attending the party were: Donald Trump, Regis Philbin, Nathan Lane and Tony Bennett.[24] In December 2010 he was honored as Turner Classic Movies Star of the Month.[25]

On February 16, 2011, Rooney was granted a temporary restraining order against Christopher Aber, one of Jan Rooney's two sons from a previous marriage.[26] On March 2, 2011 Rooney appeared before a special US Senate committee that was considering legislation to curb elder abuse. Rooney stated that he was financially abused by unnamed family members. On March 27, 2011, all of Rooney's finances were permanently handed over to lawyers over the claim of missing money.[27]

In April 2011, the temporary restraining order that Rooney was previously granted was replaced by a confidential settlement between Rooney and his stepson.[28] Christopher Aber and Jan Rooney have denied all the allegations.[29][30]

Marriages

> “Always get married early in the morning. That way, if it doesn't work out, you haven't wasted a whole day.”
>
> —Mickey Rooney[31]

Name	Years	Children
Ava Gardner	1942–1943	
Betty Jane Rase	1944–1949	Mickey Rooney, Jr. (born July 3, 1945)
		Tim Rooney (January 4, 1947 – September 23, 2006)
Martha Vickers	1949–1951	Theodore Michael Rooney (born April 13, 1950)
Elaine Devry	1952–1958	
Barbara Ann Thomason (akas: Tara Thomas, Carolyn Mitchell)	1958–1966	Kelly Ann Rooney (born September 13, 1959)
		Kerry Rooney (born December 30, 1960)
		Michael Joseph Rooney (born April 2, 1962)
		Kimmy Sue Rooney (born September 13, 1963)
Marge Lane	1966–1967	
Carolyn Hockett	1969–1975	Jimmy Rooney (adopted from Carolyn's previous marriage) (born 1966)
		Jonelle Rooney (born January 11, 1970)
Jan Chamberlin	1978–present	

Filmography

Selected films

This is a selected list of Rooney's full-length films, both theatrical and made for television.

Year	Title
1927	*Orchids and Ermine*
1932	*The Beast of the City*
	Sin's Pay Day
	High Speed
	Fast Companions
	My Pal, the King
	Officer Thirteen
1933	*The Big Cage*
	The Life of Jimmy Dolan
	The Big Chance
	Broadway to Hollywood
	The Chief
	The World Changes

1934	*Beloved*
	The Lost Jungle
	I Like It That Way
	Manhattan Melodrama
	Love Birds
	Half a Sinner
	Hide-Out
	Chained
	Blind Date
	Death on the Diamond
1935	*The County Chairman*
	Reckless
	The Healer
	A Midsummer Night's Dream
	Rendezvous
	Ah, Wilderness!
1936	*Riffraff*
	Little Lord Fauntleroy
	Down the Stretch
	The Devil is a Sissy
1937	*A Family Affair*
	Captains Courageous
	Slave Ship
	Hoosier Schoolboy
	Live, Love and Learn
	Thoroughbreds Don't Cry
	You're Only Young Once
1938	*Love Is a Headache*
	Judge Hardy's Children
	Hold That Kiss
	Lord Jeff
	Love Finds Andy Hardy
	Boys Town
	Stablemates
	Out West with the Hardys

1939	*The Adventures of Huckleberry Finn*
	The Hardys Ride High
	Andy Hardy Gets Spring Fever
	Babes in Arms
	Judge Hardy and Son
1940	*Young Tom Edison*
	Andy Hardy Meets Debutante
	Strike Up the Band
1941	*Andy Hardy's Private Secretary*
	Men of Boys Town
	Life Begins for Andy Hardy
	Babes on Broadway
1942	*The Courtship of Andy Hardy*
	A Yank at Eton
	Andy Hardy's Double Life

Year	Title
1943	*The Human Comedy*
	Thousands Cheer
	Girl Crazy
1944	*Andy Hardy's Blonde Trouble*
	National Velvet
1946	*Love Laughs at Andy Hardy*
1947	*Killer McCoy*
1948	*Summer Holiday*
	Words and Music
1949	*The Big Wheel*
1950	*Quicksand*
	The Fireball
	He's a Cockeyed Wonder
1951	*My Outlaw Brother*
	The Strip
1952	*Sound Off*
1953	*Off Limits*
	All Ashore
	A Slight Case of Larceny
1954	*Drive a Crooked Road*
	The Atomic Kid
1955	*The Bridges at Toko-Ri*
	The Twinkle in God's Eye

1956	*The Bold and the Brave*
	Francis in the Haunted House
	Magnificent Roughnecks
1957	*Operation Mad Ball*
	Baby Face Nelson
1958	*A Nice Little Bank That Should Be Robbed*
	Andy Hardy Comes Home
1959	*The Big Operator*
	The Last Mile
1960	*Platinum High School*
	The Private Lives of Adam and Eve
1961	*King of the Roaring 20's – The Story of Arnold Rothstein*
	Breakfast at Tiffany's
	Everything's Ducky
1962	*Requiem for a Heavyweight*
1963	*It's a Mad, Mad, Mad, Mad World*
1964	*The Secret Invasion*
1965	*Twenty-Four Hours to Kill*
	How to Stuff a Wild Bikini
1966	*The Devil In Love*
	Ambush Bay
1968	*Skidoo*
1969	*The Extraordinary Seaman*
	The Comic
	80 Steps to Jonah
1970	*Cockeyed Cowboys of Calico County*
	Santa Claus Is Comin' to Town (voice)
1971	*Mooch Goes to Hollywood*
	The Manipulator
1972	*Evil Roy Slade*
	Richard
	Pulp
1973	*The Godmothers*
1974	*Thunder County*
	Rachel's Man
	Journey Back to Oz (voice)
	The Year Without a Santa Claus (voice)
1975	*Ace of Hearts*
	From Hong Kong with Love

Year	Title
1976	*Find the Lady*
1977	*The Domino Principle*
	Pete's Dragon
1978	*The Magic of Lassie*
1979	*The Black Stallion*
	Arabian Adventure
	Rudolph and Frosty's Christmas in July (voice)
1981	*The Fox and the Hound* (voice)
	Bill
	Odyssey of the Pacific
1982	*The Emperor of Peru*
1983	*Bill: On His Own*
1984	*It Came Upon the Midnight Clear*
1985	*The Care Bears Movie* (voice)
1986	*Lightning, the White Stallion*
1988	*Bluegrass*
1989	*Erik the Viking*
	Little Nemo: Adventures in Slumberland (voice)
1990	*Home For Christmas*
1991	*My Heroes Have Always Been Cowboys*
1992	*The Milky Life*
	Sweet Justice
	Silent Night, Deadly Night 5: The Toy Maker
	Little Nemo: Adventures In Slumberland
	Maximum Force
1993	*The Legend of Wolf Mountain*
	The Magic Voyage (voice)
1994	*Revenge of the Red Baron*
	The Outlaws: The Legend of O.B. Taggart
	Making Waves
1995	*America: A Call to Greatness*
1997	*Killing Midnight*
1998	*The Face on the Barroom Floor*
	Animals and the Tollkeeper
	Michael Kael vs. the World News Company
	Sinbad: The Battle of the Dark Knights
	Babe: Pig in the City

1999	*Holy Hollywood*
	The First of May
2000	*Internet Love*
	Phantom of the Megaplex
2001	*Lady and the Tramp II: Scamp's Adventure* (voice)
2002	*Topa Topa Bluffs*
2003	*Paradise*
2005	*Strike the Tent*
	A Christmas Too Many
2006	*The Thirsting*
	To Kill a Mockumentary
	Night at the Museum
2007	*The Yesterday Pool*
	Bamboo Shark
2008	*Lost Stallions: The Journey Home*
	A Miser Brothers' Christmas (voice)
2010	*Gerald*
2011	*The Muppets*

Short subjects

Year	Title
1926	*Not to Be Trusted*
1927	*Mickey's Circus*
	Mickey's Pals
	Mickey's Eleven
	Mickey's Battles
1928	*Mickey's Parade*
	Mickey in School
	Mickey's Nine
	Mickey's Little Eva
	Mickey's Wild West
	Mickey in Love
	Mickey's Triumph
	Mickey's Babies
	Mickey's Movies
	Mickey's Rivals
	Mickey the Detective
	Mickey's Athletes
	Mickey's Big Game Hunt

1929	*Mickey's Great Idea*
	Mickey's Menagerie
	Mickey's Last Chance
	Mickey's Brown Derby
	Mickey's Northwest Mounted
	Mickey's Initiation
	Mickey's Midnite Follies
	Mickey's Surprise
	Mickey's Mix-Up
	Mickey's Big Moment
	Mickey's Strategy

Year	Title
1930	*Mickey's Champs*
	Mickey's Explorers
	Mickey's Master Mind
	Mickey's Luck
	Mickey's Whirlwinds
	Mickey's Warriors
	Mickey the Romeo
	Mickey's Merry Men
	Mickey's Winners
	Screen Snapshots Series 9, No. 24
	Mickey's Musketeers
	Mickey's Bargain
1931	*Mickey's Stampede*
	Mickey's Crusaders
	Mickey's Rebellion
	Mickey's Diplomacy
	Mickey's Wildcats
	Mickey's Thrill Hunters
	Mickey's Helping Hand
	Mickey's Sideline
1932	*Mickey's Busy Day*
	Mickey's Travels
	Mickey's Holiday
	Mickey's Big Business
	Mickey's Golden Rule
	Mickey's Charity

Year	Title
1933	*Mickey's Ape Man*
	Mickey's Race
	Mickey's Big Broadcast
	Mickey's Disguises
	Mickey's Touchdown
	Mickey's Tent Show
	Mickey's Covered Wagon
1934	*Mickey's Minstrels*
	Mickey's Rescue
	Mickey's Medicine Man
1935	*Pirate Party on Catalina Isle*
1937	*Cinema Circus*
1938	*Andy Hardy's Dilemma*
1940	*Rodeo Dough*
1941	*Meet the Stars #4*: Variety Reel #2
1943	*Show Business at War*
1947	*Screen Snapshots: Out of This World* Series
1953	*Screen Snapshots: Mickey Rooney – Then and Now*
1958	*Screen Snapshots: Glamorous Hollywood*
1968	*Vienna*
1974	*Just One More Time*
1975	*The Lion Roars Again*
2008	*Wreck the Halls*

Television

Rooney has made countless appearances in TV sitcoms and television films. He has also lent his voice to many animation films. Only his most important work is listed in this section.

Year(s)	Title
1954–1955	*The Mickey Rooney Show*: *Hey Mulligan*
1957	*The Comedian* (on *Playhouse 90*)
1964–1965	*Mickey*
1981	*Bill* (won Emmy, Golden Globe, and Peabody Award for role of Bill)
1982	*One of the Boys* (canceled after 13 episodes)
1983	*Bill: On His Own* (sequel to 1981's "Bill" nominated for Emmy)
1990–1993	*The Adventures of the Black Stallion*

Awards

Year	Award	Category	Nominated Work / Honor	Result
1938	Academy Award	Academy Juvenile Award	(With Deanna Durbin) "For their significant contribution in bringing to the screen the spirit and personification of youth, and as juvenile players setting a high standard of ability and achievement."	Honored
1939	Academy Award	Best Actor in a Leading Role	*Babes in Arms*	Nominated
1943	Academy Award	Best Actor in a Leading Role	*The Human Comedy*	Nominated
1956	Academy Award	Best Actor in a Supporting Role	*The Bold and the Brave*	Nominated
1957	Emmy Award	Best Single Performance in a Leading or Supporting Role	"The Comedian", episode of *Playhouse 90*	Nominated
1957	Laurel Award	Top Male Action Star	*Baby Face Nelson*	3rd Place
1958	Emmy Award	Best Single Performance	*Alcoa Theatre*	Nominated
1960	Hollywood Walk of Fame	Star of Motion Picture	Star at 1718 Vine Street	Honored
		Star of Television	Star at 6372 Hollywood Boulevard	Honored
		Star of Radio	Star at 6541 Hollywood Boulevard	Honored
1961	Emmy Award	Best Single Performance in a Leading or Supporting Role	"Somebody's Waiting", episode of *The Dick Powell Show*	Nominated
1962	Laurel Award	Top Male Supporting Performance	*Requiem for a Heavyweight*	Nominated
1964	Golden Globe	Best TV Star – Male	*Mickey*	Won
1980	Academy Award	Best Actor in a Supporting Role	*The Black Stallion*	Nominated
1981	Emmy Award	Outstanding Lead Actor in a Limited Series or Special	*Bill*	Won
1981	Golden Globe	Best Actor in a TV Mini-Series or Motion Picture	*Bill*	Won
1983	Academy Award	Academy Honorary Award	"In recognition of his 50 years of versatility in a variety of memorable film performances."	Honored
1983	Emmy Award	Outstanding Lead Actor in a Limited Series or Special	*Bill: On His Own*	Nominated
1991	Gemini Award	Best Performance by an Actor in a Continuing Leading Dramatic Role	*The Adventures of the Black Stallion*	Nominated
1991	Young Artist Award	Former Child Star Award	For lifetime achievement as a child star (Subsequently renamed "The Mickey Rooney Award")	Honored
1996	Giffoni Film Festival	François Truffaut Award	—	Honored
2004	Pocono Mountains Film Festival	Lifetime Achievement Award	—	Honored

In 1996, a Golden Palm Star on the Palm Springs, California, Walk of Stars was dedicated to him.[32]

References

[1] *Life Is Too Short.* Autobiography (1991). ISBN 978-0-679-40195-7

[2] *Current Biography 1942.* H.W. Wilson Co. (January 1942). pp704-06. ISBN 99903-960-3-5.

[3] Mickey Rooney (http://www.imdb.com/name/nm0001682/) at the Internet Movie Database

[4] Server, Lee, *Ava Gardner "Love is Nothing"* (2006), St. Martin's Press

[5] Albin, Kira. *Mickey Rooney: Hollywood, Religion and His Latest Show* (http://www.grandtimes.com/rooney.html). GrandTimes.com Senior Magazine. 1995.

[6] Gabler, Neal, *Walt Disney*, (2006), Alfred A. Knopf

[7] Rooney, Mickey. "The Lion Reigns Supreme", *MGM: When the Lion Roars*, 1992 film

[8] "11th Academy Awards" (http://www.oscars.org/awards/academyawards/legacy/ceremony/11th.html). *Oscars.org*. . Retrieved July 6, 2011.

[9] "In 1939 [Rooney] became the top box-office star in the world, a title he held for three consecutive years." Branagh, Kenneth (narrator). *1939: Hollywood's Greatest Year.* Turner Classic Movies, 2009.

[10] "12th Annual Youth in Film Awards" (http://www.youngartistawards.org/pastnoms12.htm). *YoungArtistAwards.org*. . Retrieved 2011-03-31.

[11] "13th Annual Youth in Film Awards" (http://www.youngartistawards.org/pastnoms13.htm). *YoungArtistAwards.org*. . Retrieved 2011-03-31.

[12] "23rd Annual Young Artist Awards" (http://www.youngartistawards.org/noms23A.htm). *YoungArtistAwards.org*. . Retrieved 2011-03-31.

[13] Dunning, John, *On The Air: The Encyclopedia Of Old-Time Radio* (1998), Oxford University Press

[14] Marx, Arthur (1987). *The Nine Lives of Mickey Rooney*. New York: Berkley. ISBN 978-0425105528.

[15] Marill, Alvin H. (2005). *Mickey Rooney: His Films, Television Appearances, Radio Work, Stage Shows, And Recordings* (http://books.google.com/books?id=h5WZOvH8VSUC&printsec=frontcover#v=onepage&q&f=false). Jefferson NC: McFarland. pp. 50. ISBN 0-7864-2015-4. .

[16] Brockes, Emma (October 16, 2005). "Murder in Tinseltown" (http://www.guardian.co.uk/stage/2005/oct/17/theatre). London: guardian.co.uk. . Retrieved Wednesday, July 13, 2011.

[17] Marx, Arthur, *The Nine Lives Of Mickey Rooney* (1986), Stein & Day

[18] America: A Call to Greatness at The Internet Movie Database, TV, 1995 (http://www.imdb.com/title/tt1838449/)

[19] "Come Dine With Me Celebrity Special" (http://www.channel4.com/food/on-tv/come-dine-with-me/series-7/christmas-celebrity-special-episode_p_1.html). Channel4.com. . Retrieved 2012-01-16.

[20] *Mickey Rooney makes panto debut* (http://www.channel4.com/news/articles/arts_entertainment/mickey+rooney+makes+panto+debut/1154447), December 7, 2007

[21] "Mickey Rooney: The Mickey show" (http://www.independent.co.uk/arts-entertainment/films/features/mickey-rooney-the-mickey-show-1063838.html). London: Independent.co.uk. 2008-12-14. . Retrieved 2012-01-16.

[22] "Review – Cinderella with Mickey Rooney, Milton Keynes Theatre « West End Whingers" (http://westendwhingers.wordpress.com/2009/12/06/review-cinderella-with-mickey-rooney-milton-keynes-theatre). Westendwhingers.wordpress.com. 2009-12-06. . Retrieved 2012-01-16.

[23] Mickey Rooney Grandchildren (http://www.concernpost.com/2012/03/28/top-10-celebrities-whos-married-more-than-3-times/)

[24] . "Actor Mickey Rooney Turns 90 With Upper East Side Style" (http://www.ny1.com/content/ny1_living/entertainment/125747/actor-mickey-rooney-turns-90-with-upper-east-side-style). Ny1.com. . Retrieved 2012-01-16.

[25] "Turner Classic Movies Star of the Month" (http://www.tcm.com/thismonth/article/?cid=353248). Tcm.com. 1920-09-23. . Retrieved 2012-01-16.

[26] "Mickey Rooney granted restraining order against stepson" (http://www.bbc.co.uk/news/entertainment-arts-12466486). Bbc.co.uk. 2011-02-16. . Retrieved 2012-01-16.

[27] "Mickey Rooney lawyer to control finances" (http://www.bbc.co.uk/news/entertainment-arts-12874412). Bbc.co.uk. 2011-03-27. . Retrieved 2012-01-16.

[28] "Mickey Rooney drops restraining order against stepson" (http://www.tmz.com/2011/04/06/mickey-rooney-chris-aber-restraining-order-tro-settlement-stepson-fortune-steal). Tmz.com. 2011-02-15. . Retrieved 2012-01-16.

[29] "Mickey Rooney Claims Elder Abuse: Actor's testimony to Congress helps spur bill for new crackdown" (http://www.aarp.org/relationships/caregiving/info-03-2011/mickey-rooney-claims-elder-abuse.html) by Carole Fleck and Talia Schmidt. AARP Bulletin, March 2, 2011

[30] Silverman, Stephen M. (2011-03-03). "Mickey Rooney: 'Elder Abuse Made Me Feel Trapped'" (http://www.people.com/people/article/0,,20470562,00.html). People.com. . Retrieved 2012-01-16.

[31] Mickey RooneyUS actor (1920 -). "Quote Details: Mickey Rooney: Always get married early..." (http://www.quotationspage.com/quote/23571.html). The Quotations Page. . Retrieved 2012-01-16.

[32] Palm Springs Walk of Stars by date dedicated (http://www.palmspringswalkofstars.com/web-storage/Stars/Stars dedicated by date.pdf)

Further reading

- Mickey Rooney, *Life Is Too Short* (New York: Random House, 1991)
- Arthur Marx, *The Nine Lives Of Mickey Rooney* (New York: Berkley Publishing Group, 1988 reprint)

See also

- List of surviving silent film actors

External links

- mickeyrooney.com (http://www.mickeyrooney.com), Official Site Of Mickey Rooney
- Mickey Rooney (http://www.imdb.com/name/nm1682/) at the Internet Movie Database
- Mickey Rooney (http://www.ibdb.com/person.asp?ID=58310) at the Internet Broadway Database
- Mickey Rooney (http://www.lortel.org/LLA_archive/index.cfm?search_by=people&first=Mickey&last=Rooney&middle=) at the Internet Off-Broadway Database
- Mickey Rooney on the Phil Silvers Show (http://philsilversshow.homestead.com/MickeyRooney.html)
- "Mickey Rooney on America, Christ and Judy Garland: The Hollywood Legend Speaks Out." (http://www.montrealmirror.com/ARCHIVES/1998/022698/film1.html) *Montreal Mirror* interview 1998
- Mickey Rooney (http://www.virtual-history.com/movie/person/141/mickey-rooney) at Virtual History

Big_band

Big Band	
Stylistic origins	Jazz Swing (in 1930s)
Cultural origins	1920s; United States
Typical instruments	Violins and other String instruments (1920–1934) • Saxophones • Brass instruments • Rhythm section (Guitar • Piano • Double bass • Vibes • Drums • Percussion)
Mainstream popularity	early 1930s–late 1940s Sporadic interest 50s-onward
Derivative forms	Jump blues • Philly soul • Disco (background) • Swing • Traditional pop[1] [2]
Regional scenes	
United States and United Kingdom	

A **big band** is a type of musical ensemble that originated in the United States and is associated with jazz and the Swing Era typically consisting of rhythm, brass, and woodwind instruments totaling approximately 12 to 25 musicians. The terms **jazz band**, **jazz ensemble**, **jazz orchestra**, **stage band**, **society band**, and **dance band** may describe this type of ensemble in particular contexts.

Instrumentation

A standard 17-piece instrumentation evolved in the big-bands, for which many commercial arrangements are available. This instrumentation consists of five saxophones (most often two altos, two tenors, and one baritone), four trumpets, four trombones (often including one bass trombone) and a four-piece rhythm section (composed of drums, acoustic bass or electric bass, piano and guitar).

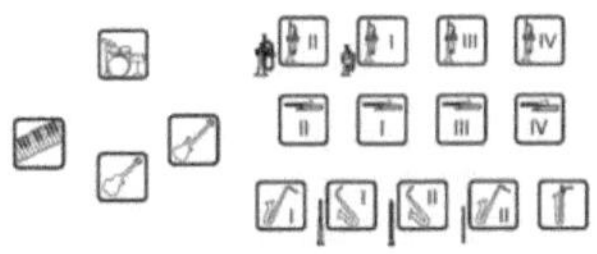

Typical seating diagram for a big band.

However, variants to this instrumentation are common. Composers, arrangers, and bandleaders have used sections with more or fewer players, and additional instruments, such as valve trombone, baritone horn/euphonium (both of which are usually used in place of or with trombones), vibes, bass clarinet, French horn, tuba, banjo, accordion and strings (violin, viola, cello). Male and female vocalists have also joined big bands to perform particular arrangements. In recent years synthesizers and / or electronic keyboards have been added, often replacing the piano.

Some arrangements call for saxophone players to double on other woodwind instruments, such as flute, clarinet, soprano sax, or bass clarinet. Trumpet and trombone players are sometimes called upon to use various sound-changing mutes, and trumpet players sometimes need to play flugelhorn. In some rhythm sections, a guitar player is omitted. Players in the rhythm section may be called upon to play acoustic or electric instruments. Latin or other auxiliary percussion instruments may be added, such as cowbells, congas, tambourines, or triangles.

History and style

There were two distinct periods in the history of popular bands. Beginning in the mid-1920s, big bands, then typically consisting of 10–25 pieces, came to dominate popular music. At that time they usually played a form of jazz that involved very little improvisation, which included a string section with violins, which was dropped after the introduction of swing in 1935. The dance form of jazz was characterized by a sweet and romantic melody. Typical of the genre were such popular artists as Paul Whiteman, Ted Lewis, Harry Reser, Leo Reisman, Abe Lyman, Nat Shilkret, George Olsen, Ben Bernie, Bob Haring, Ben Selvin, Earl Burnett, Gus Arnheim, Henry Halstead, Rudy Vallee, Jean Goldkette, Glen Gray, Isham Jones, Roger Wolfe Kahn, Sam Lanin, James Last, Vincent Lopez, Ben Pollack, Shep Fields and Fred Waring.

Paul Whiteman and his orchestra in 1921

Many of these artists changed styles or retired after the introduction of swing music. Although unashamedly commercial, these bands often featured front-rank jazz musicians—for example Paul Whiteman employed Bix Beiderbecke and Frankie Trumbauer. There were also "all-girl" bands such as "Helen Lewis and Her All-Girl Jazz Syncopators". Lewis and her band, Ben Bernie's band "Ben Bernie and All the Lads", and Roger Wolfe Kahn's band were filmed by Lee De Forest in his Phonofilm sound-on-film process in 1925, in three short films that are now in the Library of Congress film collection.

Towards the end of the 1920s, a new form of Big Band emerged which was more authentically "jazz," in that more space was given to improvised soloing. This form of music never gained the popularity of the dance form of jazz. The few recordings made in this form of jazz were labelled race records and were intended for a limited urban audience. Few white musicians were familiar with this music, Johnny Mercer, Harold Arlen and Hoagy Carmichael being notable exceptions. The three major centers in this development were New York City, Chicago and Kansas City. In the first, a sophisticated approach to arranging predominated, originally in the work of Don Redman for the Fletcher Henderson band, later in the work of Duke Ellington for his Cotton Club orchestra, and Walter 'Foots' Thomas for Cab Calloway's, Charlie Spivak and His Orchestra, and Mel Tormé's Mel-Tones. Some big ensembles, like the Joe "King" Oliver outfit played a kind of half arranged, half improvised jazz, often relying on "head" arrangements. Other great bands, like the one of Luis Russell became a vehicle for star instrumentalists, in his case Louis Armstrong. There the whole arrangement had to promote all the possibilities of the star, although they often contained very good musicians, like Henry "Red" Allen, J. C. Higginbotham and Charlie Holmes. Others such as Alvino Rey grew popular with shows in New York City and then toured the country sharing their hit songs and new musical styles.

Radio and movies

Earl "Fatha" Hines became the star of Chicago with his Grand Terrace Cafe band and began to broadcast live from The Grand Terrace nightly coast-to-coast across America. Meanwhile in Kansas City and across the Southwest, an earthier, bluesier style was developed by such bandleaders as Benny Moten and, later, by Jay McShann and Jesse Stone. Big band remotes on the major radio networks spread the music from ballrooms and clubs across the country during the 1930s and 1940s, with remote broadcasts from jazz clubs continuing into the 1950s on NBC's *Monitor*.

Radio was a major factor in gaining notice and fame for Benny Goodman, known as the "King of Swing". Soon, others challenged him, and "the battles of the bands" became a staple at theater performances featuring many groups on one bill.

Big Bands also began to appear in movies in the 1930s right on through to the 1960s. Shep Fields and his orchestra appeared in The Big Broadcast of 1938 for Paramount Pictures while accompanying the actor Bob Hope in the 1930s. Alvino Rey and His Orchestra were featured in films through RKO Pictures during their peak in the early 1940s, such as *Sing Your Worries Away*.[3] Fictionalized biographical films of Glenn Miller, Gene Krupa, Benny Goodman, and others were made in the 1950s, as nostalgic tributes to the glory years.

Rise and fall of swing

Benny Goodman and Peggy Lee

Swing music began in the 1920s, distinguished by a more supple feel than the more literal 4/4 of earlier jazz and a walking bass—Walter Page is often credited with developing this, though isolated earlier examples exist (e.g., by Wellman Braud on Ellington's *Washington Wabble* from 1927).

This type of music flourished through the early 1930s, although there was little mass audience for it until around 1936. Up until that time, it was viewed with ridicule and looked upon as a curiosity. After 1935, big bands rose to prominence playing swing music and held a major role in defining swing as a distinctive style. Western Swing musicians also formed very popular big bands during the same period.[4] [5] [6]

There was a considerable range of styles among the hundreds of popular bands. Many of the better known bands reflected the individuality of the bandleader, the lead arranger, and the personnel. Count Basie played a relaxed propulsive swing, Bob Crosby more of a dixieland style, Benny Goodman a hard driving swing, and Duke Ellington's compositions were varied and sophisticated. Many bands featured strong instrumentalists, whose sounds dominated, such as the clarinets of Benny Goodman, Artie Shaw and Woody Herman, the trombone of Jack Teagarden, the trumpet of Harry James, the drums of Gene Krupa, and the vibes of Lionel Hampton. The popularity of many of the major bands was amplified by star vocalists, such as Frank Sinatra with Tommy Dorsey, Helen O'Connell and Bob Eberly with Jimmy Dorsey, Ella Fitzgerald with Chick Webb, Billie Holiday and Jimmy Rushing with Count Basie, Dick Haymes and Helen Forrest with Harry James, Doris Day with Les Brown, Toni Arden and Ken Curtis with Shep Fields and Peggy Lee with Benny Goodman. Some bands were **society bands** that relied on strong ensembles but little on soloists or vocalists, such as the bands of Guy Lombardo and Paul Whiteman.

By this time the big band was such a dominant force in jazz that the older generation found they either had to adapt to it or simply retire—with no market for small-group recordings (made worse by a depression-era industry reluctant to take risks), some musicians such as Louis Armstrong and Earl Hines fronted their own bands, while others, like Jelly Roll Morton and King Oliver, lapsed into obscurity.

The major African American bands of the 1930s included, apart from the bands led by Ellington, Hines and Calloway, were those of Count Basie, Jimmie Lunceford, and Chick Webb.

Benny Goodman, Artie Shaw, Tommy Dorsey, Shep Fields and, later, Glenn Miller far eclipsed other bands in popularity from the middle of the decade. Also popular was the Casa Loma Orchestra and Benny Goodman's early band.

During World War II Earl Hines and Billy Eckstine led bands whose soloists figured as the early performers of Bebop. For a few years after the war, Dizzy Gillespie and Lionel Hampton led Bebop-oriented big bands.

Glenn Miller

White teenagers and young adults were the principal fans of the Big Bands in the late 1930s and early 1940s. They danced to recordings and the radio, and attended live concerts whenever they could. They were knowledgeable and often biased toward their favorite bands and songs, and sometimes worshipful of the famous soloists and vocalists. Many bands toured the country in grueling one-night stands to reach out to their fans. Traveling conditions and lodging were often difficult, in part due to segregation in most parts of the United States, and the personnel often had to perform on little sleep and food. Apart from the star soloists, many personnel received low wages and would abandon the tour and go home if bookings fell through. Personal problems and intra-band discord could affect the playing of the group. Drinking and addictions were common. Turnover was frequent in many bands, and top soloists were often lured away to better contracts. Sometimes bandstands were too small, public address systems inadequate, pianos out of tune. Successful bandleaders dealt with all these hazards of touring to hold their bands together—some with rigid discipline (Glenn Miller), some with canny psychology (Duke Ellington).

Big Bands played a major role in lifting morale during World War II. Many band members served in the military and toured with USO troupes at the front, with Glenn Miller losing his life while traveling between troop shows. Many bands suffered from the loss of personnel and quality declined at home during the war years. A recording ban from an ill-timed recording strike in 1942 worsened the situation. Vocalists began to strike out on their own and by the end of the war, swing was giving way to less danceable music including bebop. Many of the great swing bands broke up as tastes changed.

Since 1945

As jazz evolved and expanded in new directions, major band performances of note did occur from the 1950s to the 1970s. Noteworthy performers included: Dizzy Gillespie, Gene Krupa, Buddy Rich, Gil Evans, Stan Kenton, Johnny Richards, Sun Ra, Gary MacFarland, Charles Mingus, Oliver Nelson, Carla Bley, Thad Jones/Mel Lewis Big Band, Sam Rivers, Don Ellis, Toshiko Akiyoshi – Lew Tabackin Big Band, Lincoln Center Jazz Orchestra and Anthony Braxton.

Ockbrook Big Band at Pride Park Stadium

Later bandleaders pioneered the performance of various Brazilian and Afro-Cuban styles with the traditional big band instrumentation, and big bands led by arranger Gil Evans, saxophonist John Coltrane (on the album *Ascension* from 1965) and electric bassist Jaco Pastorius introduced cool jazz, free jazz and jazz fusion, respectively, to the big band domain. Modern big bands can be found playing all styles of jazz music. Some large contemporary European jazz ensembles play mostly avant-garde jazz using the instrumentation of the big bands. Examples include the Vienna Art Orchestra, founded in 1977, and the Italian Instabile Orchestra, active in the 1990s. In the late 1990s, swing made a comeback in the US. The Lindy Hop has taken hold on both coasts, and many younger people took an interest in big band styles again. The Jazz at Lincoln Center Orchestra with Wynton Marsalis is the resident orchestra of Jazz at Lincoln Center (JALC). The JALC Orchestra currently tours internationally, promoting the big band sound.

African "Afrobeat" big bands have existed from 1970 to the present when Fela Kuti of Nigeria, fused big band jazz with Yoruba tribal rhythms, highlife, and American James Brown funk music. As of 2008 there are over 40 working afrobeat big bands [7] including Antibalas, Chicago Afrobeat Project, Chopteeth, Femi Kuti, and Seun Kuti.

Big Band Arrangements

In contrast to jazz "combos," in which musical performances are largely improvised, big band music is primarily crafted in advance by an arranger.

Typical big band arrangements of the swing period are written in strophic form with the same phrase and chord structure repeated several times. Each iteration, or *chorus*, most commonly follows Twelve bar blues form or Thirty-two-bar (AABA) song form. The first chorus of an arrangement typically introduces the melody, and is followed by subsequent choruses of development. This development may take the form of improvised solos, written *soli* sections, and *shout choruses*.

An arrangement's first chorus is sometimes preceded by an introduction, which may be as short as a few measures or may extend to chorus of its own. Many arrangements contain an interlude, often similar in content to the introduction, inserted between some or all choruses. Other methods of embellishing the form include modulations and cadential extensions.

See also

- Big band remote
- British dance band
- List of American big band bandleaders
- List of big bands
- List of British big band leaders
- List of experimental big bands
- Swing (jazz performance style)
- Territory band

References

[1] http://www.allmusic.com/explore/style/d29
[2] Pop standards
[3] Sing Your Worries Away information (http://www.imdb.com/title/tt0035338/)
[4] (http://www.cmt.com/artists/az/wills_bob/bio.jhtml)
[5] (http://www.cmt.com/artists/az/cooley_spade/bio.jhtml)
[6] (http://www.cmt.com/artists/az/williams_tex/bio.jhtml)

Further reading

- William Russo, *Composing for the Jazz Orchestra* University of Chicago Press, Library of Congress no. 61-8642
- George T. Simon, *The Big Bands*, The Macmillan Company, New York, 1967, Library of Congress no. 67-26643

Sayonara

WARNING: Article could not be rendered - ouputting plain text.
Potential causes of the problem are: (a) a bug in the pdf-writer software (b) problematic Mediawiki markup (c) table is too wide

Sayonaraoriginal movie posterDirected byJoshua LoganProduced byWilliam GoetzWritten byJames Michener (novel)Paul OsbornStarringMarlon BrandoPatricia Owens (actress)Patricia OwensJames GarnerMartha ScottMiiko TakaMiyoshi UmekiRed ButtonsRicardo MontalbanEditing byArthur P. SchmidtPhilip W. Anderson (editor)Philip W. AndersonDistributed byWarner Bros.Release date(s) December 5, 1957Running time 147 minutesCountry United StatesLanguage EnglishJapaneseBox office $22,000,115 (in U.S.)Sayonara is a 1957 color (Technicolor) American film starring Marlon Brando. It tells the story of an United StatesAmerican United States Air ForceAir Force flier who was an "ace" fighter pilot during the Korean War.Sayonara won four Academy Awards, including acting honors for co-stars Red Buttons and Miyoshi Umeki.The film's screenplay was adapted by Paul Osborn from the Sayonara (novel)novel by James Michener, and it was produced by William Goetz and directed by Joshua Logan. Unlike most 1950s romantic dramas, Sayonara deals squarely with racism and prejudice.Shales, Tom (July 14, 2006). "The Bright Appeal of Red Buttons". The Washington Post. . Retrieved May 23, 2010.Plot Lloyd "Ace" Gruver, a major and the son of a U.S. Army general, is stationed at Itami Air Force Base (now Osaka International Airport) near Kobe, Japan. He falls in love with a Japanese entertainer (Hana-ogi) who is a performer for a Takarazuka RevueTakarazuka-like theater company, whom he meets through his enlisted crew chief, Airman Kelly. Kelly is about to wed a Japanese woman, Katsumi, in spite of the disapproval of the United States military, which will not recognize the marriage. The Air Force, including Gruver, is against the marriage. Gruver and Kelly have an argument during which Gruver uses a racial slur to describe Kelly's fiancee. Gruver eventually apologizes, then agrees to be Kelly's best man at the wedding. Kelly suffers further prejudice at the hands of a particularly nasty colonel, pulling extra duty and all the less-attractive assignments. When he and many others who are married to Japanese are ordered back to the States, Kelly realizes he will not be able to take his wife, who is now pregnant. Finding no other way to be together, Kelly and Katsumi commit double suicide. This strengthens Major Gruver's resolve to marry his own Japanese lover. When asked by a Stars and Stripes (newspaper)Stars and Stripes reporter what will he say to both the "big brass" as well as to the Japanese, neither of which will be particularly happy, Major Gruver says, "Tell 'em we said 'Sayonara.'"The ending in the movie, however, differs from that of the book, in which Gruver says "sayonara" to his Japanese girlfriend and returns to the States. CastMarlon Brando - Maj. Lloyd "Ace" Gruver, USAF Patricia Owens (actress)Patricia Owens - Eileen Webster James Garner - Capt. Mike Bailey, USMC Martha Scott - Mrs. Webster Miiko Taka - Hana-ogi Miyoshi Umeki - Katsumi Kelly Red Buttons - Airman Joe Kelly Kent Smith - Lt. Gen. Mark Webster Reiko Kuba - Fumiko Soo Yong - Teruko Ricardo Montalban - Nakamura Production Brando adopted a nondescript American EnglishSouthern accent for Gruver, despite the objections of director Logan, who did not think a Southern accent was appropriate for a general's son who was educated at West Point. Logan later admitted to the author and journalist Truman Capote about Brando, "I've never worked with such an exciting, inventive actor. So pliable. He takes direction beautifully, and yet he always has something to add. He's made up this Southern accent for the part; I never would have thought of it myself, but, well, it's exactly right — it's perfection."Capote, Truman (2008), Portraits and Observations, New York: Modern Library, p. 191Critical receptionSayonara has received widespread critical acclaim, particularly for its writing and cinematography, in addition to the acting ability of its cast. It has won four Academy Awards, including acting honors for co-stars Red Buttons and Miyoshi Umeki. Review aggregator Rotten Tomatoes reports that 100% critics have given the film a positive review, with a rating average of 7.2/10.The film earned $10.5 million in rentals in North America."All Time Domestic Champs", Variety, 6 January 1960 p 34Legacy Alongside the less successful Japanese War Bride and The Teahouse of the August Moon (film)The Teahouse of the August Moon, Sayonara was argued by some scholars to have increased racial tolerance in the United States by openly discussing interracial marriages.Sarah Kovner (2012). Occupying Power: Sex Workers and Servicemen in Postwar Japan. Stanford University Press. pp. 65–66. ISBN 978-0-8047-8346-0. . Other scholars have argued that the movie is one in a long list stereotyping Asian American women as "lotus blossom, geisha girl, china doll, or The World of Suzie WongSuzie Wong" by presenting Asian women as "passive, sexually compliant and easy to seduce" or as downright prostitutes.Edith Wen-Chu Chen (2010). Encyclopedia of Asian American Issues Today. ABC-CLIO. pp. 644–645. ISBN 978-0-313-34751-1. .AwardsSayonara won multiple Academy Awards for "The 30th Academy Awards (1958) Nominees and Winners". oscars.org. . Retrieved 2011-08-21. "NY Times: Sayonara". NY Times. . Retrieved 2008-12-22.Academy Award for Best Supporting ActorBest Actor in a Supporting Role (Red Buttons) Academy Award for Best Supporting ActressBest Actress in a Supporting Role (Miyoshi Umeki) Academy Award for Best Art DirectionBest Art Direction-Set Decoration (Ted Haworth & Robert Priestley) Academy Award for SoundBest Sound (George Groves (sound engineer)George Groves) It was also nominated for Academy Award for Best ActorBest Actor in a Leading Role (Marlon Brando) Academy Award for Best CinematographyBest Cinematography (Ellsworth Fredericks) Academy Award for DirectingBest Director (Joshua Logan) Academy Award for Film EditingBest Film Editing (Arthur P. Schmidt & Philip W. Anderson) Academy Award for Best PictureBest Picture (William Goetz) Academy Award for Writing Adapted ScreenplayBest Writing, Screenplay Based on Material from Another Medium (Paul Osborn) ReferencesExternal links Sayonara at the Internet Movie Database Sayonara at AllRovi Sayonara at the Turner Classic MoviesTCM Movie Database Sayonara at Rotten Tomatoes Trailer of Sayonara introduced by Miika Taka The Duke and His Domain by Truman Capote James Garner Interview on the Charlie Rose Show James Garner interview at Archive of American Television - (c/o Google Video) - March 17, 1999

Miyoshi_Umeki

Miyoshi Umeki	

Born	Umeki MiyoshiMay 8, 1929Otaru, Japan
Died	August 28, 2007 (aged 78)Licking, Missouri, U.S.
Occupation	Actress/Singer
Years active	1953–1972
Spouse(s)	Wynn Opie (1958–1967) Randall Hood (1968–1976)

Miyoshi Umeki (*Umeki Miyoshi*, o r *Miyoshi Umeki*; May 8, 1929 – August 28, 2007[1]) was a naturalized American actress and standards singer. She was best known for her roles as Katsumi, the wife of Joe Kelly (Red Buttons), in the 1957 film *Sayonara*, as Mei Li in the 1958 Broadway musical and 1961 film *Flower Drum Song*, and as Mrs. Livingston, the housekeeper of Bill Bixby's and Brandon Cruz's characters, in the TV series *The Courtship of Eddie's Father*. She was a shin Issei or post-1945 immigrant from Japan. She was a Tony Award- and Golden Globe-nominated actress and was the first and only Asian woman to win an Academy Award for acting.[2]

Life

Born as **Umeki Miyoshi** (in Japanese, a person's family name is placed before their given name) in Otaru, on the Japanese island of Hokkaidō,[1] she was the youngest of nine children. Her father owned an iron factory.[1] After World War II, Umeki began her career as a nightclub singer in Japan, using the name **Nancy Umeki**,[3] Her early influences were traditional Kabuki theater and American pop music.[1] Later, in one of her appearances on *The Merv Griffin Show*, she treated viewers to her impression of singer Billy Eckstine, one of her American favorites growing up.

Career

She recorded for RCA Victor Japan[1] from 1950–1954 and appeared in the film *Seishun Jazu Musume*. She recorded mostly American jazz standards, which she sang partially in Japanese and partially in English, or solely in either language. Some of the songs she sang during this period were It Isn't Fair, Sentimental Me, My Foolish Heart, With A Song In My Heart, Again, Vaya Con Dios, (How Much Is) That Doggie in the Window? and I'll Walk Alone. She moved to the United States in 1955[1] [3] and after appearing on the *Arthur Godfrey Talent Scouts* (she was a series regular for one season),[4] she signed with the Mercury Records label and released several singles and two albums.[1]

Her appearances on the Godfrey program brought her to the attention of director Joshua Logan, who cast her in *Sayonara*.[2] In 1958, she appeared twice on the NBC variety show, *The Gisele MacKenzie Show* in which she performed "How Deep Is the Ocean". Umeki won an Academy Award for Best Supporting Actress for her role in *Sayonara*. She was the first Asian performer to win an Academy Award for acting .[2]

In 1958, she was nominated for a Tony Award for Best Leading Actress in a Musical for her performance in the Broadway premiere production of the musical *Flower Drum Song*,[2] where she played Mei-Li.[5] The show ran for two years. A *Time* magazine cover story remarked that *"[T]he warmth of her art works a kind of tranquil magic"*.[1] Umeki went on to appear in the film adaptation of the musical.[3] She was nominated for a Golden Globe Award for *Flower Drum Song*.

Although a guest on many television variety shows, she appeared in only four more motion pictures through 1962, including the film version of *Flower Drum Song* (1961). The others were *Cry for Happy* (1961), *The Horizontal Lieutenant* (1962) and *A Girl Named Tamiko* (1963). From 1969-1972 she appeared in *The Courtship of Eddie's Father* as Mrs. Livingston, the housekeeper, for which she was again nominated for a Golden Globe Award. She retired from acting following the end of the series.

Personal life

Her first marriage, to television director Frederick Winfield "Wynn" Opie in 1958, ended in divorce[1] in 1967. The couple had one son — Michael H. Opie, born in 1964.[1] She married Randall Hood in 1968, and her son's name was changed to Michael Randall Hood.[6] The couple operated a Los Angeles-based business renting editing equipment to film studios and university film programs.[1] Randall Hood died in 1976.[2]

Death

According to her son, Umeki lived in Sherman Oaks for a number of years [3] before moving to Licking, Missouri, to be near her son and his family, which included two grandchildren. She died there of complications from cancer.

Discography

RCA Records Japan (1950-1954)

During her recording career in Japan, Miyoshi recorded the following songs:

- Sleepy My Love (1950)
- Under The Moonlight (1950)
- Don't Say That Person's Name (1950)
- Evening Whisper (1950)
- I Feel Like Crying (1950)
- I'm Waiting For You (1950)
- One Night Of Sorrow (1951)
- Misery (1951)
- It Isn't Fair (1951)
- Sentimental Me (1951)
- My Foolish Heart (1953)
- Why Won't You Believe Me (live) (1953)
- Again (1953)
- Manhattan Moon (1953)
- With A Song In My Heart (1953)
- I'll Walk Alone (1953)
- My Baby's Coming Home (1953)
- Silent Night (1953)

- I'm Walking Behind You (1953)
- (How Much Is) That Doggie in the Window? (1953)
- Sayonara (the Japanese Farewell Song) (1953)
- My Ichiban Tomodachi (1953)
- Vaya Con Dios (1954)
- Kiss Me Again Stranger (1954)
- My Ichiban Tomodachi (live) (1954)
- Sayonara (the Japanese Farewell Song) (live) (1954)

(two other Japanese language songs were recorded in 1952)

Singles on Mercury Records (1955-1959)

Miyoshi signed with Mercury Records in 1955 and recorded the following 45 rpm singles:

- How Deep Is The Ocean/Why Talk (1955)
- The Little Lost Dog/The Story You're About To Hear Is True (1956)
- The Mountain Beyond The Moon/Oh What Good Company We Could Be(with Red Buttons) (1957)
- Sayonara(The Japanese Farewell Song)/Be Sweet Tonight (1957)
- Sayonara/On And On (1957)

Miyoshi recorded a version of Pick Yourself Up for Mercury Records in 1959, but the song was never released.

Albums on Mercury Records

Miyoshi Sings For Arthur Godfrey(MG-20165)(1956) Tracks:

- If I Give My Heart To You
- China Nights (Shina no yoru)
- I'm In The Mood For Love
- My Baby's Coming Home
- How Deep Is The Ocean
- Slowly Go Out of Your Mind
- Teach Me Tonight
- Hanna Ko San
- Can't Help Loving That Man
- S'Wonderful
- Over The Rainbow
- Sayonara (The Japanese Farewell Song)

Miyoshi (MG-20568) (1959) Tracks:

- My Heart Stood Still
- My Ship
- You Make Me Feel So Young
- They Can't Take That Away From Me
- Sometimes I'm Happy
- I'm Old Fashioned
- That Old Feeling (song)
- Gone With The Wind (song)
- Jeepers Creepers
- Wonder Why
- I Could Write A Book

Miyoshi - Singing Star of Rodgers And Hammerstein's Flower Drum Song (MGW-12148)(1958) (this is a reissue of the Arthur Godfrey album with some tracks replaced) Tracks:

- Sayonara
- If I Give My Heart To You
- China Nights (Shina no yoru)
- I'm In The Mood For Love
- My Baby's Coming Home
- How Deep Is The Ocean
- Slowly Go Out Of Your Mind
- Teach Me Tonight
- Hanna Ko San
- Can't Help Loving That Man
- Over The Rainbow
- The Little Lost Dog

Film theme recordings

Miyoshi Umeki recorded the theme song for the film *Sayonara* in 1957. She recorded the title song for the film *Cry For Happy* in which she starred in 1961.

- *Flower Drum Song* (Broadway Original Cast) (1958) (Sony Records)

Tracks by Miyoshi Umeki:

- A Hundred Million Miracles
- I Am Going To Like It Here
- Don't Marry Me
- Wedding Parade/A Hundred Million Miracles

- *Flower Drum Song* (Film Soundtrack) (1961) (Decca Records)

Miyoshi Umeki sings the same songs on both the film soundtrack and the Broadway show

Filmography

Film

Year	Title	Role	Notes
1953	*Seishun Jazz musume* (Seishun jazu musume)	Kashu	
1956	*Around the World Revue*	Nancy Umeki	also known as Universal Musical Short 2655: Around the World Revue
1957	*Sayonara*	Katsumi	• Academy Award for Best Supporting Actress • Nominated - Golden Globe Award for Best Supporting Actress - Motion Picture
1961	*Flower Drum Song*	Mei Li	Nominated — Golden Globe Award for Best Actress - Motion Picture Musical or Comedy
1961	*Cry for Happy*	Harue	
1962	*A Girl Named Tamiko*	Eiko	
1962	*The Horizontal Lieutenant*	Akiko	

Television

Year	Title	Role	Notes
1955	*Arthur Godfrey and His Friends*	Herself	Regular performer
1957	*The Perry Como Show*	Herself	1 episode
1958–61	*The Dinah Shore Chevy Show*	Herself	Episode #2.32 (1958), Episode #4.16 (1960), Episode #5.17
1958	*What's My Line?*	Herself - Mystery Guest	Episode #414 dated 11 May 1958
1958	*The Tennessee Ernie Ford Show*	Herself	Episode #2.25
1959	*The Chevy Showroom Starring Andy Williams*	Herself	Episode #2.2
1959	*Toast of the Town*	Singer	
1961	*Here's Hollywood*	Herself	Episode dated 27 December 1961
1961–62	*The Donna Reed Show*	Kimi	2 episodes: "The Geisha Girl" (1961), "Aloha, Kimi" (1962)
1962	*The Andy Williams Show*	Herself	Episode dated 11 October 1962, Episode dated 13 December 1962
1962	*Hallmark Hall of Fame*	Lotus-Blossom	"The Teahouse of the August Moon"
1962	*Sam Benedict*	Sumiko Matsui	"Tears for a Nobody Doll"
1963	*Rawhide*	Nami	"Incident of the Geisha"
1963	*Dr. Kildare*	Hana Shigera	"One Clear Bright Thursday Morning"
1964	*Burke's Law*	Mary 'Lotus Bud' Ling	"Who Killed the Paper Dragon?"
1964	*The Virginian*	Kim Ho	"Smile of a Dragon"
1964	*Mister Ed*	Ako Tenaka	"Ed in the Peace Corps"
1964	*The Celebrity Game*	Herself	Episode dated 19 April 1964
1969	*The Queen and I*	Japanese Bride	"The Trousseau"
1969–72	*The Courtship of Eddie's Father*	Mrs. Livingston	• 66 episodes • Nominated — Golden Globe Award for Best Supporting Actress – Series, Miniseries or Television Film
1971	*This Is Your Life*	Herself	for Bill Bixby
1971	*The Pet Set*	Herself	Episode dated 30 June 1971
1971	*The Merv Griffin Show*	Herself	Episode dated 29 March 1971
1972	*Salute to Oscar Hammerstein II*	Herself	

References

[1] Bernstein, Adam. "Actress Miyoshi Umeki, 78, Dies of Cancer" (http://www.washingtonpost.com/wp-dyn/content/article/2007/09/05/AR2007090501484.html). *The Washington Post*. 5 September 2007,

[2] "Oscar winner Miyoshi Umeki dies at 78" (http://www.usatoday.com/life/people/2007-09-05-umeki-obit_N.htm). *USA Today*. 5 September 2007.

[3] Lavietes, Stuart. "Miyoshi Umeki, Oscar-winning actress, dies at 78" (http://www.iht.com/articles/2007/09/06/america/obits.php). *International Herald Tribune*. 6 September 2007.

[4] Miyoshi Umeki (http://www.imdb.com/name/nm0880855/) at the Internet Movie Database

[5] Miyoshi Umeki (http://www.ibdb.com/person.asp?ID=62909) at the Internet Broadway Database

[6] *Miyoshi Umeki, first Asian to win an Oscar, dies* (http://afp.google.com/article/ALeqM5go1cOxEboLMN38e03MYjlMjJ9y4A). AFP. 6 September 2007.

External links

- Miyoshi Umeki (http://www.imdb.com/name/nm0880855/) at the Internet Movie Database
- Miyoshi Umeki (http://www.ibdb.com/person.asp?ID=62909) at the Internet Broadway Database
- *New York Times* bio (http://movies2.nytimes.com/gst/movies/filmography.html?p_id=72422&mod=bio)
- Miyoshi Umeki (http://www.findagrave.com/cgi-bin/fg.cgi?page=gr&GRid=21382379) at *Find a Grave*

Hatari!

WARNING: Article could not be rendered - ouputting plain text.

Potential causes of the problem are: (a) a bug in the pdf-writer software (b) problematic Mediawiki markup (c) table is too wide

Hatari!original movie poster by Frank McCarthy (artist)Frank McCarthyDirected byHoward HawksProduced byHoward HawksPaul HelmickWritten byHarry KurnitzLeigh BrackettStarringJohn WayneElsa MartinelliHardy KrügerRed ButtonsMusic byHenry ManciniCinematographyRussell HarlanDistributed byParamount PicturesRelease date(s) October 6, 1961Running time 157 minCountry United StatesLanguage EnglishBox office $7,000,000 (US/ Canada)"All-time top film grossers", Variety 8 January 1964 p 37. Please note this figure is rentals accruing to film distributors not total money earned at the box office.. Hatari! (pronounced [hɑtɑri], Swahili languageSwahili for "Danger!") is a 1962 United StatesAmerican film directed by Howard Hawks and starring John Wayne. Portraying a group of hunters, the film includes dramatic wildlife chases and the magnificent backdrop scenery of Mount Meru (Tanzania)Mount Meru, a dormant volcano.Hatari! was filmed on location in Tanganyika (in what is now northern Tanzania).The film gathers its several characters from different parts of the world: Sean Mercer (John Wayne, USA), Pockets (Red Buttons, USA), Anna Maria 'Dallas' D'Alessandro (Elsa Martinelli, Italy), Kurt Müller (Hardy Krüger, Germany), Brandy De la Court (Michele Girardon, France), Charles 'Chips' Maurey (Gerard Blain, France), Luis Francisco Garcia Lopez (Valentin de Vargas, Mexico). Plot The story involves a group of Western expatriates catching wild animals in East Africa and selling them to zoos, led by Sean Mercer (John Wayne). At the outset, Little Wolf (Bruce Cabot), aka "The Indian," is injured during a rhinoceros chase. He is rushed to the hospital at Arusha, where he needs a blood transfusion. None of the team can provide the required rare blood type, but a young Frenchman, Charles "Chips" Maury (Gerard Blain) can. Chips is hired as a replacement for The Indian as the international group tries to fill all its animal orders for the season. A wildlife photographer, Anna Maria D'Alessandro (Elsa Martinelli), arrives to take photos of the animals caught by Mercer's men for a zoo that will be buying the bulk of that season's captures. She is at first mistaken for a man because of the introduction letter she had signed with only her initials. She soon acquires the nickname "Dallas." During her stay, Dallas also becomes known as Mama Tembo (Mother of Elephants) for her efforts to save three baby elephants, culminating in a chase through the streets of Arusha. She also develops a crush on Mercer. There are romantic subplots between the other characters as well. CastJohn Wayne as Sean Mercer Hardy Krüger as Kurt Müller Elsa Martinelli as Anna Maria 'Dallas' D'Alessandro Red Buttons as Pockets Gérard Blain as Charles 'Chips' Maurey Bruce Cabot as Little Wolf aka The Indian Michèle Girardon as Brandy de la Court Valentin de Vargas as Luis Francisco Garcia Lopez Eduard Franz as Dr. Sanderson Queenie Leonard as Nurse (scenes deleted) ProductionHatari! has a very loose script and, like many other major works of Hawks, is principally structured on the relationships among the characters, though it is 'bookended' by the initial violent (and nearly fatal) encounter with a rhinoceros and the end-of-season determination to make such a capture to fulfill the team's quota. Much of the film revolves around scenes of chasing animals in jeeps and trucks across the African plains. The animals pursued are also all live, wild, and untrained, a procedure banned today over concerns of exhausting and killing the targeted animals. The script was written by Hawks' favorite writer, Leigh Brackett, after the group returned from Africa with the hunting scenes.As the animals frequently refused to make noise 'on cue' (in particular, the baby elephants refused to trumpet inside populated areas), local Arusha game experts and zoo collectors were hired to do 'animal voice impersonations'. Hawks stated in interviews that he had originally planned to star both Clark Gable and Wayne in the film until Gable's death finally ruled that out. Hatari! introduced the memorable Henry Mancini tune "Baby Elephant Walk". Other memorable musical moments involve a duet of Stephen Foster's "Old Folks at Home" (Swanee River) with Martinelli's character playing the piano, and Red Buttons' character playing the harmonica. Also, after 'the Indian' is left to recuperate in the hospital, the members of the group get drunk and sing a somewhat tragicomical song that features lyrics like "Oh whisky leave me alone, I'm tired and I want to go home".According to director Howard Hawks, all the animal captures in the picture were actually performed by the actors; no stuntmen or animal handlers were substituted onscreen. The rhino really did escape, and the actors really did have to recapture it - and Hawks included the sequence for its realism. Much of the action sequence audio had to be re-dubbed due to John Wayne's cursing while wrestling with the animals. However, a stand-in, "rusty" Walkley, was used for some scenes involving Elsa Martinelli.Australian Womans Weekly December 5, 1962 at http://trove.nla.gov.au/ndp/del/article/44555225?searchTerm=GIRAFFES%20CAN%20DANCE%20&searchLimits=See alsoJohn Wayne filmographyReferencesExternal links Hatari! at the Internet Movie Database hatari in the Internet Living Swahili Dictionary Hatari Lodge: A History

The Longest Day (film)

WARNING: Article could not be rendered - ouputting plain text.

Potential causes of the problem are: (a) a bug in the pdf-writer software (b) problematic Mediawiki markup (c) table is too wide

The Longest Dayoriginal movie posterDirected byKen Annakin (British & French exteriors)Andrew Marton (American exteriors)Bernhard Wicki (German episodes)Gerd Oswald (parachute drop)Darryl F. Zanuck (uncredited)Produced by Darryl F. ZanuckScreenplay byRomain GaryJames Jones (author)James JonesDavid Pursall Cornelius RyanJack Seddon Based onThe Longest Day (book)The Longest Day by Cornelius RyanStarringJohn WayneHenry FondaRobert MitchumSean ConneryEddie AlbertCurd JürgensRichard BurtonPeter LawfordRod SteigerIrina DemickGert FröbeEdmond O'BrienKenneth MoreMusic byMaurice JarreCinematographyJean Bourgoin Walter Wottitz Editing by Samuel E. BeetleyDistributed by20th Century FoxRelease date(s)25 September 1962 (France)4 October (US)23 October (UK)Running time 178 minutesCountry United StatesLanguageEnglish languageEnglish (British & American scenes)German languageGerman (German scenes)French languageFrench (French scenes)Budget US$10,000,000 (est.) or $7.75 millionSolomon, Aubrey. Twentieth Century Fox: A Corporate and Financial History (The Scarecrow Filmmakers Series). Lanham, Maryland: Scarecrow Press, 1989. ISBN 978-0-8108-4244-1. p253Box office US$50,000,000 (worldwide, to 12/1963)The Longest Day is a 1962 war film based on the 1959 in literature1959 history book The Longest Day (book)The Longest Day by Cornelius Ryan, about D-Day, the Normandy landings on 6 June 1944, during World War II.Producer Darryl F. Zanuck paid the author of the book, Cornelius Ryan, USD$175,000 for the screen rights to produce the film."Operation Overblown". - Time (magazine)TIME. - October 19, 1962. - Retrieved: 2008-06-23 The film was adapted from the book by Romain Gary, James Jones (author)James Jones, David Pursall, Jack Seddon, and the author himself. It was directed by Ken Annakin (British and French exteriors), Andrew Marton (American exteriors), Gerd Oswald (parachute drop scene), Bernhard Wicki (German scenes), and Darryl F. Zanuck (uncredited). Many of the military consultants and advisors who helped with the film's production were actual participants in the action on D-Day, and are portrayed in the film. The producers drew them from both sides; Allied and Axis. Among them are Günther Blumentritt (a former German general), James M. Gavin (an American general), Frederick E. MorganFrederick Morgan (Deputy Chief of Staff at Supreme Headquarters Allied Expeditionary ForceSHAEF), John Howard (British Army officer)John Howard (who led the airborne assault on the Pegasus Bridge), Simon Fraser, 15th Lord LovatLord Lovat (who commanded the 1st Special Service Brigade), Philippe Kieffer (who led his men in the assault on Ouistreham), Marie Pierre KoenigPierre Koenig (who commanded the Free French Forces in the invasion), Max-Josef PemselMax Pemsel (a German general), Werner Pluskat (the major who was the first German officer to see the invasion fleet), Josef PrillerJosef "Pips" Priller (the hot-headed pilot) and Lucie Rommel (widow of Erwin Rommel). Unique for British and American produced World War II films at the time, all French and German characters speak in their own languages with subtitles in English languageEnglish. A separate version exists, shot simultaneously, in which all the actors speak their lines in English, which is why the Trailer (film)trailer has the Germans delivering their lines in English. This version saw limited use during the initial release, and more extensive use during a late 1960s re-release of the film. The English-only version was featured on the "flip side" of an older single disc DVD release. The film, one of the very few 1960s epics made in black and white, features a large ensemble cast including actors such as John Wayne, Kenneth More, Richard Todd (who took part in the actual invasion; his battalion reinforced John Howard's Pegasus Force), Robert Mitchum, Richard Burton, Sean Connery, Henry Fonda, Red Buttons, Rod Steiger, Leo Genn, Peter Lawford, Gert Fröbe, Irina Demick, Bourvil, Curd Jürgens, Robert Wagner, Paul Anka and Arletty.

Several of these actors played roles that were virtually cameo appearances. The film was shot on French locations including the Île de Ré, Saleccia beach in Saint-Florent, Haute-Corse, Port-en-Bessin-Huppain filling in for Ouistreham, Les Studios de Boulogne in Boulogne-Billancourt and the actual locations of Pegasus Bridge near Bénouville, Calvados, Sainte-Mere-Eglise and Pointe du Hoc.Editors of La Presse de la Manche Notre Jour Plus Long 2012CastAmericans Actor Role Eddie AlbertColonel Thompson, 29th Infantry Division (United States)29th Infantry DivisionPaul AnkaUnited States Army RangersU.S. Army RangerRichard BeymerPrivate (rank)Private Arthur 'Dutch' Schultz, 82nd Airborne DivisionRed ButtonsJohn Steele (paratrooper)John Steele, 82nd Airborne Division Mark DamonPrivate (rank)Private Harris Ray DantonCaptain (OF-2)Captain Frank Fred DurU.S. Army Ranger MajorFabian (entertainer)Fabian ForteU.S. Army RangerMel FerrerMajor General Robert Haines Henry FondaTheodore Roosevelt Jr.Brigadier General Theodore Roosevelt Jr., Deputy Commander U.S. 4th Infantry Division4th Infantry Div.Steve Forrest (actor)Steve ForrestCaptain (U.S. Army)Captain Harding, 82nd Airborne Division Henry GraceDwight D. EisenhowerGeneral Dwight D. Eisenhower, SHAEFSupreme Allied CommanderPeter Helm Young GI (military)GIJeffrey HunterSergeant (later LieutenantLt.) John H. Fuller Alexander KnoxWalter Bedell SmithLieutenant General Walter Bedell Smith, SHAEF Chief of Staff Dewey Martin (actor)Dewey MartinPrivate (rank)Private Wilder (role cut from released version) Roddy McDowallPrivate (rank)Private Morris, 4th Infantry Division (United States)4th Infantry DivisionJohn MeillonAlan Goodrich KirkAdmiral Alan G. Kirk, Senior U.S. Naval Commander Sal MineoPrivate (rank)Private Martini Robert MitchumNorman CotaBrigadier General Norman Cota, Asst. Commander U.S. 29th Infantry Division29th Infantry Div.Edmond O'BrienRaymond O. BartonMajor General Raymond O. Barton, Commander 4th Infantry Division (United States)4th Infantry Div.Ron Randell Joe Williams Robert RyanJames M. GavinBrigadier General James M. Gavin, Asst. Commander 82nd Airborne Division82nd Airborne Div.Tommy SandsU.S. Army RangerGeorge SegalU.S. Army RangerRod SteigerDestroyer Commander Nicholas Stuart Omar BradleyLieutenant General Omar N. Bradley, Commander First United States ArmyU.S. First ArmyTom TryonLieutenant Wilson, 82nd Airborne Division Robert WagnerU.S. Army RangerJohn WayneBenjamin H. VandervoortLt. Col. Benjamin Vandervoort,Commander 2nd Battalion, 505th Infantry Regiment (United States)505th Parachute Infantry Reg.Stuart WhitmanLieutenant Sheen, 82nd Airborne Division British Actor Role Patrick BarrGroup Captain James Martin StaggJ.M. StaggRichard BurtonFlying Officer David Campbell Bryan Coleman Ronald Callen Sean Connery Private Flanagan Leslie de Laspee Private Bill Millin, No. 4 Commando (Piper on Beach) Frank Finlay Private Coke, 2nd Oxfordshire and Buckinghamshire Light InfantryOxford & Bucks L.I.Harry Fowler Private, 2nd Oxford & Bucks L.I. Leo GennBrigadier Edwin P. Parker Jr. Harold Goodwin (English actor)Harold Goodwin Private, 2nd Oxford & Bucks L.I. John Gregson British Padre 6th Airborne Division (United Kingdom)6th Airborne DivisionJack Hedley RAF Briefing Officer Donald HoustonRoyal Air ForceRAF pilot at flight base Simon LackTrafford Leigh-MalloryAir Chief Marshal Trafford Leigh-Mallory, Commander Allied Air Forces Peter Lawford Brigadier Simon Fraser, 15th Lord LovatLord Lovat, Commander 1st Special Service Brigade Victor Maddern Private Hutchinson, 3rd Infantry Division Howard Marion-Crawford Glider Doctor Michael Medwin Private Watney, 3rd Infantry Division Kenneth More Capt. Colin Maud Royal Navy beach master Louis MounierArthur Tedder, 1st Baron TedderAir Marshal Arthur William Tedder, SHAEFDeputy Supreme Allied CommanderLeslie Phillips Royal Air Force officer Trevor ReidBernard L. MontgomeryGeneral Bernard Montgomery, SHAEFCommander Allied Ground ForcesJohn Robinson (actor)John RobinsonBertram RamsayAdmiral Bertram Ramsay, SHAEFCommander Allied Naval ForcesNorman Rossington Private Clough Richard ToddMajor John Howard (British army officer)John Howard, CO 2nd Oxford & Bucks L.I. Richard Wattis British Paratrooper officer, 6th Airborne Division French Actor Role Arletty Madame Barrault Jean-Louis BarraultPriestFather Louis Roulland BourvilAndré Bourvil Mayor of Colleville-sur-MerCollevillePauline Carton Maid Gil Delamare Naval Commando (also was the leading stunt director of the film) Irina Demick Janine Boitard (French Resistance)Fernand Ledoux Louis Christian MarquandPhilippe KiefferCapitaine de Frégate Philippe KiefferCommander Naval commandos (France)French Navy commandosMadeleine Renaud Mother Superior Georges RivièreSergent Guy de MontlaurJean ServaisCounter AdmiralContre-amiral Jaujard Georges Wilson

Alexandre Renaud Germans Actor Role Hans Christian BlechMajor (Germany)Major Werner Pluskat, 352nd Infantry Division (Germany)352nd Infantry DivisionWolfgang BüttnerGeneralleutnant Hans SpeidelDr. Hans Speidel, chief of staff, Army Group BGert FröbeUnteroffizier "Kaffeekanne" ("coffee pot") Paul Hartmann (actor)Paul HartmannGerd von RundstedtGeneralfeldmarschall Gerd von Rundstedt, commander OB West Werner HinzErwin RommelGeneralfeldmarschall Erwin Rommel, commander Army Group B Karl John (actor)Karl JohnGeneralleutnant Wolfgang Häger Curd Jürgens General der Infanterie Günther Blumentritt, chief of staff, OB WestTil KiweHauptmann Helmuth Lang, Rommel's aide Wolfgang Lukschy Generaloberst Alfred Jodl, chief of staff, OKW Kurt Meisel Ernst Düring Richard MünchGeneral der Artillerie Erich Marcks, commander LXXXIV Army Corps Hartmut Reck Bernhard Bergsdorf Heinz ReinckeOberst Josef Priller, commander Jagdgeschwader 26JG 26Ernst Schröder (actor)Ernst SchröderGeneraloberst Hans von Salmuth, commander 15th Army (Wehrmacht)15th ArmyHeinz Spitzner Helmuth Meyer Wolfgang Preiss Generalmajor Max-Josef PemselMax Pemsel chief of intelligence, 7th Army (Wehrmacht)7th ArmyPeter van EyckOberstleutnant Ocker, Pluskat's Commanding Officer Vicco von BülowVicco "Loriot" von Bülow Luftwaffe Chief's aid. CastingJohn Wayne in The Longest Day The role of Lieutenant Colonel Benjamin H. Vandervoort was actively sought by Charlton Heston, but the last-minute decision of John Wayne to take a role in the film prevented Heston from participating. While everyone else accepted twenty-five thousand dollars as payment, John Wayne insisted on a quarter of a million dollars to punish Zanuck for referring to him as "poor John Wayne" regarding Wayne's problems with his lavish movie The Alamo (1960 film)The Alamo.Wills, Garry (1997). John Wayne's America: The Politics of Celebrity. Simon & Schuster. ISBN 978-0-684-80823-9. Sergeant Kaffeekanne's name is German languageGerman for "coffee pot", which he always carries. It is a common misconception that Bill Millin, the piper who accompanies Simon Christopher Joseph Fraser, 17th Lord LovatLord Lovat to Normandy with his bagpipes, played himself in the film. He was actually portrayed by Pipe Major Leslie de Laspee, the official piper to the Queen Mother in 1961. "Piper Bill Millin". The Pegasus Archive. . Retrieved 2007-11-01. "D-Day Piper - Bill Millin". The Miniatures Page. 2006-08-03. . Retrieved 2007-11-01. In Sainte-Mère-Église, John Steele (paratrooper)Private John Steele from the 82nd Airborne (played by Red Buttons) has been memorialized by the local population with a dummy hanging from a parachute from the church tower on which he accidentally landed. Richard Todd, who played Major John Howard (British Army officer)John Howard, leader of the British Airborne assault on the Pegasus Bridge, took part in the real bridge assault on D-Day. Todd was offered the chance to play himself, but thought the part would be too small, so he asked to play the part of Major John Howard instead. Shortly after the British have captured the Orne bridge (later renamed Horsa Bridge) in the film, one of the soldiers tells Todd, playing Howard, that all they have to do now is sit tight and wait for the 7th Parachute Battalion to relieve them, to which Todd replies dismissively that the Paras are always late. This was a private joke, as Todd had been the adjutant of the 7th Parachute Battalion on D-Day and did help relieve the forces on the bridge. During the scene of Todd as Howard awaiting relief from the beachhead, an officer in a Para beret next to Todd, playing Howard, is an actor playing Richard Todd. Joseph Lowe landed on Omaha Beach and scaled the cliffs at Pointe du Hoc on D-Day. He repeated the climb for the cameras 17 years later as a serving member of the 505th Infantry Regiment (United States)505th Airborne Battle Group who provided US Army film extras. Actor Curd Jürgens portrayed the German General Blumentritt who muses on the incompetence of his superiors. Jürgens himself was actually imprisoned by the Nazis in his youth. Former U.S. President Dwight D. Eisenhower was considered for the role of himself in the film, and he indicated his willingness. However, it was decided that makeup artists couldn't make him appear young enough to play his World War II self. The role of General Eisenhower went to Henry Grace, a set decorator with no acting experience, but who had been in the film industry since the mid-1930s. He was a dead ringer for the younger Eisenhower, though his voice differed. Acclaimed British actor Christopher Lee auditioned for a role, but was turned down because he did not look like a military man, even though he had served in the Royal Air Force as an Intelligence Officer. However, some film books incorrectly credit him with a role in the film. The film was the last one made by Sean Connery before he was cast in the role of James Bond. It is also noticeable that James Bond's later rivals, "Auric Goldfinger" and "Karl Stromberg", are played by the GermanyGerman actors Gert Fröbe and Curd Jürgens respectively, who coincidentally also take "Nemesis" roles in this movie (Sergeant Kaffeekanne, and General Günther Blumentritt). Canadian journalist Charles Lynch (journalist)Charles Lynch played an unnamed war correspondent on Juno beach, who calls the homing pigeons "damned traitors" when they fly towards Germany. Lynch, then working for Reuters, was one of a handful of journalists to accompany the troops ashore in the actual invasion. Sean Connery had a part in both movies made from Cornelius Ryan's non-fiction books: Private Flanagan in The Longest Day; Major General Urquhart in A Bridge too Far. Wolfgang Preiss similarly had parts in the same two films: Maj. Gen. Max Pemsel in The Longest Day; Field Marshal Von Rundstedt in A Bridge too Far. Mel Ferrer was originally signed to play the role of General James M. Gavin but withdrew from the role due to a scheduling conflict.Notre jour le plus long La Presse de la Manche 2012Filming During the filming of the landings at Omaha Beach, the extras appearing as American soldiers did not want to jump off the landing craft into the water because they thought it would be too cold. Robert Mitchum, who played General Norman Cota, became disgusted with their trepidation. He jumped in first, at which point the extras followed his example. The paradummyRupert paradummies used in the film were far more elaborate and lifelike than those actually used for the decoy parachute drop (Operation Titanic), which were actually just canvas or burlap sacks filled with sand. In the real operation, six Special Air Service soldiers jumped with the dummies and played recordings of loud battle noises to distract the Germans. At $10,000,000, this film was the most expensive black-and-white film made until 1993, when Schindler's List was released. In the scenes where the paratroopers land, the background noise of frogs croaking "ribbit ribbit" was wrong for northern French frog species and showed that the film probably used an American recording of background night noises. Colin Maud loaned Kenneth More the Shillelagh (club)shillelagh he carried ashore in the actual invasion, while Richard Todd wore the actual D-Day helmet worn by John Howard (British Army officer)Major John Howard. In the film, 3 Free French Special Air Service paratroopers jumped into France before British and American airborne landings. This is accurate. 36 Free French SAS (4 sticks) jumped into Brittany (Plumelec and Duault) on 5 June (11 h 30). The first Allied soldier killed in action was Free French Corporal Émile Bouétard: 6 June (0 h 40) in Plumelec, Morbihan. The United States Sixth Fleet extensively supported the filming and made available many amphibious landing ships and craft for scenes filmed in Corsica, though many of the ships were of (then) modern vintage. The USS SpringfieldUSS Springfield, and USS Little RockUSS Little Rock, both WW2 light cruisers (though extensively reconfigured into guided missile cruisers) were used in the shore bombardment scenes, though it was easy to tell they did not resemble their wartime configuration. Among historical inaccuracies was the calling up of Engineers to "blow a path" through a "sea wall" to allow forward troop movement. The Sea Wall was, at highest, one meter, made of rocks, and lay at the edge of high tide. This barrier was used to prevent erosion and on June 6, 1944, provided cover for the overwhelmed first two waves.Omama Beach, June 6, 1944, Joseph BaloskiAwardsAcademy Awards for Best Art Direction (1962): Ted Haworth, Léon Barsacq, Vincent Korda and Gabriel Béchir (nominated) "The Longest Day (1962) Awards". Turner Classic Movies, A Time Warner Company. . Retrieved 2008-04-30.Academy Awards for Best Cinematography (1962): Jean Bourgoin and Walter Wottitz (won)Academy Awards for Best Editing (1962): Samuel E. Beetley (nominated)Academy Awards for Best Picture (1962): (nominated)Academy Awards for Best Special Effects (1962): (won)See alsoJohn Wayne filmographyReferencesExternal links The Longest Day at the Internet Movie Database The Longest Day at the Turner Classic MoviesTCM Movie Database The Longest Day at AllRovi The Longest Headache

The Eleventh Hour (1962 TV series)

WARNING: Article could not be rendered - ouputting plain text.
Potential causes of the problem are: (a) a bug in the pdf-writer software (b) problematic Mediawiki markup (c) table is too wide

The Eleventh HourGenreMedical dramaStarringWendell CoreyJack GingRalph BellamyTheme music composer C. King PalmerHarry SukmanOpening theme "The Film Opens"Composer(s) Harry SukmanCountry of origin United StatesLanguage(s) EnglishNo. of seasons 2No. of episodes 62ProductionExecutive producer(s)Norman FeltonProducer(s)Sam RolfeCamera setupMulti-camera setupMulti-cameraRunning time 45–48 minProduction company(s) Arena ProductionsMGM TelevisionBroadcastOriginal channelNBCPicture formatBlack-and-whiteAudio formatMonauralOriginal run October 3, 1962 – April 22, 1964The Eleventh Hour is an American medical drama about psychiatry starring Wendell Corey, Jack Ging, and Ralph Bellamy, which aired 62 new episodes plus selected rebroadcasts on NBC from October 3, 1962, to September 9, 1964.Series premise The series, loosely comparable to the 1961 NBC hit Dr. Kildare, starring Richard Chamberlain and Raymond Massey, reveals the human stories of people who come to the psychiatrist either through private practice, a hospital, or a court of law. In 1963, the series shared a two-part crossover episode with Dr. Kildare; both programs used the theme of wise teacher and young intern.The term "eleventh hour" refers to a time of last resort in an aggrieved person's life, as he faces a potential nervous breakdown. Ging appeared in both seasons as Dr. Paul Graham, a clinical psychologist to Corey's first-season character of the psychiatrist Dr. Theodore Bassett, advisor to the Department of Corrections. The first season hence offered episodes about the mental health of criminals. In the second season, which ended new episodes on April 22, 1964, Bellamy replaced Corey in the role of Dr. Richard Starke, a psychiatrist engaged in private practice. The executive producer was Norman Felton; Sam Rolfe was the Television productionproducer.Alex McNeil, Total Television, New York: Penguin Books, 1996, 4th ed., p. 255Guest starsPhilip AbbottNeile AdamsEddie AlbertLola AlbrightFrank AletterRichard AndersonEdward AndrewsEdward AsnerFrankie AvalonPhyllis AveryMartin BalsamJoanna BarnesHerschel BernardiCharles BickfordBill BixbyBeau BridgesLloyd BridgesLloyd BochnerRichard Bull (actor)Richard BullPaul BurkeRed ButtonsJames T. CallahanJoseph CampanellaMary Grace CanfieldDiahann CarrollVeronica CartwrightLinden ChilesJames CoburnMichael ConstantineNoreen CorcoranPatricia CrowleyKim DarbyColleen DewhurstBradford DillmanElinor DonahueTony DowHoward DuffDan DuryeaAndrew DugganKeir DulleaLinda EvansShelley FabaresFabian (entertainer)FabianNorman FellAnne FrancisJames FranciscusBeverly GarlandHarold GouldDon Gordon (actor)Don GordonDon GradyDabbs GreerVirginia GreggJames Gregory (actor)James GregoryHarry GuardinoEileen HeckartAnne HelmPeter HelmCheryl HoldridgeCeleste HolmRon HowardKim HunterDiana HylandDavid JanssenHenry Jones (actor)Henry JonesKaty JuradoNoah KeenShirley KnightTed KnightHarvey KormanBert LahrElsa LanchesterRobert Lansing (actor)Robert LansingAngela LansburyPiper LaurieBethel LeslieJoanne LinvilleRobert LoggiaJulie LondonLynn LoringJames MacArthurRoddy McDowallJohn McGiverBarbara McNairScott MarloweWalter MatthauJayne MeadowsBurgess MeredithDina MerrillVera MilesElizabeth MontgomeryBill MumyEd NelsonLois NettletonLeonard NimoyJeanette NolanEdmond O'BrienCarroll O'ConnorJerry ParisEleanor ParkerMichael ParksRoger PerryCliff RobertsonRuth RomanMarion RossBarbara RushKurt RussellRobert RyanAlbert SalmiTelly SavalasGeorge C. ScottJacqueline ScottSylvia SidneyTom SimcoxJean StapletonInger StevensDean StockwellMaxine StuartKarl SwensonRoy ThinnesJoan TompkinsFranchot ToneRip TornHarry TownesMaxine StuartRobert Wagner (actor)Robert WagnerTuesday WeldFaye WrayKeenan WynnScheduling notesThe Eleventh Hour aired on Wednesday evenings following Perry Como's Kraft Music Hall and Espionage (TV series)Espionage. In its first season, The Eleventh Hour was placed opposite the alternating anthology series, Armstrong Circle Theatre and The United States Steel Hour on CBS and the last season of Naked City (TV series)Naked City on American Broadcasting CompanyABC. In the second season, The Eleventh Hour faced The Danny Kaye Show variety program on CBS and Channing (TV series)Channing on ABC.1962-1963; 1963-1964 American network television schedule; from appendix of Total TelevisionSee alsoBreaking Point (1963 TV series)Breaking Point, a similar TV show to this one. ReferencesExternal links The Eleventh Hour at the Internet Movie Database The Eleventh Hour at TV.com

Little House on the Prairie (TV series)

WARNING: Article could not be rendered - ouputting plain text.
Potential causes of the problem are: (a) a bug in the pdf-writer software (b) problematic Mediawiki markup (c) table is too wide

Little House on the PrairieGenreDramatic programmingFamily DramaWestern (genre)WesternDeveloped byBlanche HanalisDirected by William F. ClaxtonMaury DexterVictor FrenchMichael LandonLeo PennStarringMichael LandonMelissa GilbertKaren GrassleMelissa Sue AndersonLindsay and Sidney GreenbushMatthew LaborteauxRichard Bull (actor)Richard BullKatherine MacGregorAlison ArngrimJonathan GilbertKevin HagenDabbs GreerVictor FrenchMerlin OlsenDean Butler (actor)Dean ButlerLinwood BoomerTheme music composerDavid Rose (musician)David RoseComposer(s) David RoseCountry of origin United StatesLanguage(s) EnglishNo. of seasons 9No. of episodes 203 (+ 5 specials) (List of Little House on the Prairie episodesList of episodes)ProductionExecutive producer(s) Michael LandonEd FriendlyProducer(s) John HawkinsWilliam F. ClaxtonRunning time 45–48 minutesProduction company(s) Ed Friendly ProductionsUniversal TelevisionNBC ProductionsDistributorWorldvision EnterprisesCBS Television Distribution (USA)NBCUniversal Television DistributionNBCUniversal International Television DistributionMGM Worldwide TelevisionMGM International Television Distribution (non-USA)BroadcastOriginal channelNBCPicture formatNTSCAudio formatMonauralOriginal run September 11, 1974 – March 21, 1983ChronologyPreceded byLittle House on the Prairie (film)Little House on the Prairie (film)Followed byLittle House on the Prairie: A Look Back to YesterdayRelated showsFather MurphyLittle House on the Prairie is an American Western (genre)Western Dramatic programmingdrama television series, starring Michael Landon, Melissa Gilbert, and Karen Grassle, about a family living on a farm in Walnut Grove, Minnesota, in the 1870s and 1880s. The show is an adaptation of Laura Ingalls Wilder's best-selling series of Little House on the PrairieLittle House books. Television producer and NBC executive Ed Friendly became aware of the story in the early 1970s. He asked Michael Landon to direct the pilot movie, who agreed on the condition that he could also play Charles Ingalls.The regular series was preceded by the Little House on the Prairie (film)two-hour pilot movie, which first aired on March 30, 1974. The series began on the NBC network on September 11, 1974, and ended on May 10, 1982. During the 1982-83 television season, with the departure of Michael Landon and Karen Grassle, the series was broadcast with the new title Little House: A New Beginning.In 1997, TV Guide ranked the two-part episode "I'll Be Waving As You Drive Away" #97 on its 100 Greatest Episodes of All Time."Special Collectors' Issue: 100 Greatest Episodes of All Time". TV Guide (June 28-July 4). 1997.Detail Although it differed from the original books and many new characters and situations were added, this television series was one of the few long-running successful dramatic family shows (and it is still in syndication). Although predominantly a drama, the program did have some comedic moments, thanks to supporting cast members Mr. Edwards (played by Victor French) and the Oleson family: Nels Oleson (Richard Bull (actor)Richard Bull), Harriet Oleson (Katherine MacGregor), Willie Oleson (Jonathan Gilbert), Nellie Oleson (Alison Arngrim), and their adopted child, Nancy Oleson (Allison Balson). The show's central characters are Charles Ingalls (Michael Landon), farmer and patriarch, with his wife, Caroline (Karen Grassle), and four daughters, Mary (Melissa Sue Anderson), Laura (Melissa Gilbert), Carrie (Lindsay and Sidney Greenbush) and Grace (Wendi and Brenda Turnbaugh), later adding adopted children Albert (Matthew Laborteaux), Cassandra (Missy Francis) and James (Jason Bateman). Other essential characters included the friendly Nels Oleson, proprietor of the town's general store, Oleson's Mercantile; his malicious, gossiping wife, Harriet; and their two spoiled children, Nellie and Willie; and later, their adopted child, Nancy (Allison Balson). Also appearing in the series are former professional football player Merlin Olsen (as Jonathan Garvey), Dabbs Greer (as Reverend Robert Alden), Karl Swenson (as Lars Hanson, the town's founder and proprietor of the town's mill), and Kevin Hagen (as Dr. Hiram Baker, the town's physician). Malcolm in the Middle creator Linwood Boomer appears as Mary Ingalls's teacher-turned-husband, Adam Kendall, whom she meets at the school for the blind in the 1978-1979 season. In the sixth season, Dean Butler (actor)Dean Butler joined the cast as Almanzo Wilder, and he and Laura are married in the seventh season premiere.Michael Landon directed the largest number of episodes (87); producer William F. Claxton handled the majority of the remaining shows (68). Co-star Victor French helmed 19 episodes, and Maury Dexter directed a handful. The series theme song was titled The Little House and was written and conducted by David Rose (musician)David Rose. The ending theme music, also written by Rose, originally appeared as a piece of incidental music in a later-season episode of Michael Landon's previous long-running series, Bonanza.As with most TV series set in a distant time or place, historical inaccuracies and errors do appear. At the end of the episode "Country Girls" an airplane can be heard flying overhead. Little House on the Prairie was largely filmed on Big Sky Ranch at Simi Valley, California and Sonora, California; as a result camera vistas sometimes picked up rugged chaparral terrain, far too mountainous and scrubby for Minnesota, and one episode even depicted Laura running away and climbing up a mountain. California's oak savannas appears in many of the scenes and are considered to be representative of the real Walnut Grove. Dr. Baker's telephone was an anachronism since the telephone existed only in large cities in the 1880s. Also during the series run, several married women take teaching jobs during an era when only single women could teach. Several episodes mentioned peanut butter sandwiches, which were not introduced until the early 1900s. Another episode ("Wave Of The Future", from season eight) shows an elderly man, ostensibly Colonel Harland Sanders, attempting to sell his restaurant franchise to Mrs. Oleson. Sanders was actually born in 1890. In the episode "A Wiser Heart", Laura attends an 1885 lecture by Ralph Waldo Emerson; Emerson actually died in 1882. The show also used the Flag of MinnesotaMinnesota flag, which was not adopted until 1957.Themes Little House explored many themes, and every episode was filled with family values, love, friendship, and faith. Adoption, alcoholism, racism and blindness were portrayed in the scripts. Some plots also included subjects such as drug addiction (i.e. morphine), leukemia, prejudice, and even rape.Several of the episodes written by Michael Landon were recycled storylines from ones that he had written for Bonanza. Season two's "A Matter of Faith" was based on the Bonanza episode "A Matter of Circumstance"; season five's "Someone Please Love Me" was based on the Bonanza episode "A Dream To Dream"; season seven's "The Silent Cry" was based on the Bonanza episode "The Sound of Sadness"; season eight's "He Was Only Twelve" was based on the Bonanza episode "He Was Only Seven"; and season nine's "Little Lou" was based on the Bonanza episode "It's A Small World."Cast and charactersMelissa Gilbert as Laura Ingalls, 1975 Melissa Gilbert has the most appearances of the series. She was absent for 13 episodes, for a total of 190 of the 203 episodes. Michael Landon appeared in all but four episodes of seasons one through eight, but departed from the cast when the show was retooled as Little House: A New Beginning (season nine). For episode counts of all cast members, see: List of Little House on the Prairie (TV series) charactersList of Little House on the Prairie characters.Main castMichael Landon as Charles Ingalls (1974–83) Karen Grassle as Caroline IngallsCaroline Quiner Ingalls (1974–82) Melissa Gilbert as Laura Ingalls Wilder (1974–83) Melissa Sue Anderson as Mary IngallsMary Ingalls Kendall (1974–81) Lindsay and Sidney Greenbush as Carrie Ingalls (1974–82) Richard Bull (actor)Richard Bull as Nels Oleson (1974–83) Katherine MacGregor as Harriet Oleson (1974–83) Alison Arngrim as Nellie Oleson Dalton (1974–81, 1982) Jonathan Gilbert as Willie Oleson (1974–83) Victor French as Isaiah Edwards (1974–77, 1979, 1981–83) Bonnie Bartlett as Grace Snider Edwards (1974–77, 1979) Kevin Hagen as Dr. Hiram Baker (1974–83) Dabbs Greer as Robert AldenRev. Robert Alden (1974–83) Charlotte Stewart as Eva Beadle Simms (1974–78) Karl Swenson as Lars Hanson (1974–78) Radames Pera as John (Sanderson, Jr.) Edwards (1975–77) Brian Part as Carl (Sanderson) Edwards (1975–77) Kyle Richards as Alicia (Sanderson) Edwards (1975–77, 1979, 1982) Merlin Olsen as Jonathan Garvey (1977–81) Hersha Parady as Alice Garvey (1977–80) Patrick LabyorteauxPatrick Laborteaux as Andrew "Andy" Garvey (1977–81) Matthew Laborteaux as Albert Quinn IngallsAlbert (Quinn) Ingalls (1978–82, 1983) Linwood Boomer as Adam Kendall (1978–81) Ketty Lester as Hester-Sue Terhune (1978–83) Wendi and Brenda Turnbaugh as Grace Ingalls (1978–82) Dean Butler (actor)Dean

Butler as Almanzo Wilder (1979–83) Lucy Lee Flippin as Eliza Jane Wilder (1979–82) Steve Tracy as Percival Dalton (1980–81, 1982) Allison Balson as Nancy Oleson (1981–83) Jason Bateman as James (Cooper) Ingalls (1981–82) Missy Francis as Cassandra (Cooper) Ingalls (1981–82) Jennifer and Michele Steffin as Rose Wilder LaneRose Wilder (1982–83) Shannen Doherty as Jenny Wilder (1982–83) Guest stars Many actors who guest-starred on the show went on to become famous; among these are: Michael Conrad and James SikkingJames B. Sikking of Hill Street Blues, Willie Aames, Anne Archer, Jonathan Banks, Billy Barty, Todd Bridges, Joshua Bryant, Matt Clark (actor)Matt Clark, James Cromwell, David Faustino, Gil Gerard, Louis Gossett, Jr., Mariette Hartley, John Hillerman, Rance Howard, Ernie Hudson, Rick Hurst, Robert Loggia, Mike Lookinland, Chuck McCann, Richard Mulligan, Sean Penn (uncredited extra), Anne Ramsey, Madeleine Stowe, and Kim Richards. Other notable guests include: Johnny CashJohnny and June Carter Cash, Landon's Bonanza co-star Mitch Vogel, and two of Landon's children Michael Landon, Jr. and Leslie Landon.Spin-offs and sequels Little House: A New Beginning When Michael Landon decided to leave the show (though he stayed on as executive producer and occasional writer and director), season nine was renamed, the focus was put on the characters of Laura and Almanzo, and more characters were added to the cast. A new family, the Carters (Stan Ivar as John, Pamela Roylance as Sarah, Lindsay Kennedy as older son Jeb, and David Friedman (actor)David Friedman as younger son Jason), move into the Ingalls' home. Meanwhile, Almanzo and Laura take in their niece, Jenny Wilder (Shannen Doherty), when Almanzo's brother dies and raise her alongside their daughter, Rose. The Wilders appear prominently in some episodes, while in others they appear only in early scenes used to introduce the story or its characters. The explanation given for the original characters' absence was that they moved to Burr Oak, Iowa, to build a much better life. The show lost viewers, and this version of the series was cancelled after nineteen episodes. However, the show lived on for two more years in movie format. Movie specials Three made-for-television post-series movies followed: Little House on the Prairie: A Look Back to Yesterday (1983), Little House: Bless All the Dear Children (1983), and Little House: The Last Farewell (1984). In The Last Farewell, Charles and Caroline decide to visit Walnut Grove. They learn that a railroad tycoon actually holds the deed to the township, and he wants to take it over for his own financial gain. Despite their best efforts, the townspeople are unable to drive the businessman away. At a town meeting, one of the residents offers a supply of explosives that he has. Each man takes turn blowing up his own building. "The Last Farewell Summary". CBS Interactive Inc.. . Retrieved 17 July 2010. When asked why the set was blown up, the show's producer, Kent McCray, said that when the series started, he made an agreement with the property owners that at the end of the series he would put the acreage back to its original state. When the production crew were estimating the cost of dismantling all the buildings, Michael Landon thought for awhile and said, "What if we blow up the town? That would get the buildings all in pieces and you can bring in your equipment to pick up the debris and cart it away." He then said that he would write it where they blow up all the buildings, except for the little house and the church. Both McCray and Landon wept as the town blew up.Parker, Lennon. "The REAL story of the destruction of Walnut Grove". Prairie Fans. . Retrieved November 14, 2009.Two other Little House movies were made in conjunction with the Landon series: the 1974 pilot for the program and Little House Years (1979), a Thanksgiving (United States)Thanksgiving special/clip show that aired in the middle of season six.Broadcast history For the first two seasons, the show was aired on Wednesday nights at 8pm ET/7pm CT, to moderate ratings. In 1976, the series became a Monday night staple on NBC; after the move, it remained in the top 30 for the rest of its run. Nielsen ratingsSeason 1 (1974–75): #13 Season 2 (1975–76): #16 Season 3 (1976–77): #7 Season 4 (1977–78): #14 Season 5 (1978–79): #16 Season 6 (1979–80): #10 Season 7 (1980–81): #25 Season 8 (1981–82): #29 Season 9 (1982–83): #28 "TV Ratings". Classic TV Hits. .Awards1976: TP de Oro, Mejor Actriz Extranjera (Best Foreign Actress), Karen Grassle1976: TP de Oro, Mejor Serie Extranjera (Best Foreign Series)1978: Emmy Award for Outstanding Cinematography in Entertainment Programming for a Series, Ted Voightlander, episode "The Fighter" 1979: Emmy Award for Outstanding Cinematography for a Series, Ted Voightlander, episode "The Craftsman" 1979: Emmy Award for Outstanding Music Composition for a Series, David Rose (musician)David Rose, episode "The Craftsman" 1980: TP de Oro, Mejor Actriz Extranjera (Best Foreign Actress), Melissa Sue Anderson1981: Western Writers of America Spur Award for Best TV Script, Michael Landon, episode "May We Make Them

Proud" 1982: Emmy Award for Outstanding Achievement in Music Composition for a Series (Dramatic Underscore), David Rose, episode "He Was Only Twelve", part 2 1983: Young Artist Award for Best Young Actress in a Drama Series, Melissa Gilbert1984: Young Artist Award for Best Young Actress in a Drama Series, Melissa Gilbert Syndication The show is popular in reruns in syndication. In the U.S., in addition to airing on local stations, it currently airs on Hallmark Channel and The Inspiration NetworkINSP. In the past, it has aired on TV Land and TBS (TV channel)TBS, where the show had a successful 20-year run. Because of its historical context and its connection to the book series, it is deemed acceptable for use by the Federal Communications CommissionFCC to meet federal E/I programming guidelines. The show is typically stripping (television)stripped (run five days a week) in syndication, which is enough to completely cover a TV station's E/I requirements and more. In Canada, reruns of the series began airing weeknights on Crossroads Television SystemCTS, a Christian-based network, as of September 1, 2008. In the US, television syndication rights are currently owned by CBS Television Distribution. Originally, NBC licensed these rights to Worldvision Enterprises, since networks could not own syndication arms at the time. As a result of corporate changes, Paramount Domestic Television and CBS Paramount Domestic Television would inherit the rights, finally passing to CTD in 2007. NBC owns ancillary rights and thus is the worldwide licensor for DVD rights as well. Sister company NBC Universal Television GroupNBC Universal International Television distributes the series internationally. DVD releases The entire series has been released on DVD. The North American DVD sets include interviews by former cast members Patrick Laborteaux and actors Alison Arngrim, Dabbs Greer and Dean Butler. A majority of the episodes in the North American DVD versions have scenes cut from the episodes—these are derived from the syndicated television versions by Worldvision Enterprises, the series former distributor. Other episodes are time-compressed; these are NTSC-converted video prints from UK PAL masters. Only a handful of episodes in the DVD sets are in their original uncut versions (for example, many season one episodes on DVD contain scenes not in current syndication prints). Many episodes on the DVD versions contain tracking lines and audio problems. Edited episodes, like the ones on Region 1 DVD, are usually about 35–38 minutes, while full and complete episodes are 44–48 minutes. The DVD sets sold in the U.S. and Canada were released under license from NBC Universal by Imavision Distribution, a company based in Quebec. Imavision has also released a French-language version of the DVD set. Both versions are in NTSC color and coded for all regions. Later copies were distributed by Lions Gate Entertainment following their acquisition of Imavision. The DVD sets sold in the United Kingdom were released by Universal Playback (a Universal Studios Home Entertainment label); this version is in PAL color and coded for region 2. Some time earlier, some single Little House episodes were released on both DVD and VHS by GoodTimes Entertainment. Before retail DVDs were available, the Little House episodes were available through a Columbia House club subscription. These VHS tapes contained two episodes per tape and were only available at a club price. The episodes on these VHS tapes, unlike the current DVDs, were not edited and remain the only commercially available uncut episodes.In November 2008, the Finnish Board of Film Classification rated the DVD release of the Little House on the Prairie series as suitable for adults only—requiring a sticker to be affixed to all DVDs saying "Banned for under-18s". This was because of Universal Pictures' decision not to submit the series to state review to avoid the state review fee of approximately USD $27,500.Suoninen, Sakari (November 5, 2008). "Finland rates "Little House on Prairie" adults-only". Reuters. .Season Episodes Originally aired DVD release date Region 1Region 2 1 23 1974–1975 July 8, 2003 July 25, 2005 2 22 1975–1976 July 8, 2003 March 27, 2006 3 21 1976–1977 November 4, 2003 March 10, 2008 4 22 1977–1978 February 17, 2004 May 26, 2008 5 24 1978–1979 June 29, 2004 August 4, 2008 6 24 1979–1980 October 26, 2004 May 3, 2010 7 24 1980–1981 February 15, 2005 July 17, 2010 8 22 1981–1982 June 14, 2005 March 20, 2011 9 19 1982–1983 November 1, 2005 January 20, 2012 3-Movie Box Set 3 TV Movies 1983–1984 November 28, 2006 TBA The Complete Television Series 203 1974–1984 November 11, 2008 TBA ReferencesExternal links Little House on the Prairie at the Internet Movie Database Little House on the Prairie at TV.com Little House: Look Back to Yesterday at the Internet Movie Database Little House: The Last Farewell at the Internet Movie Database Little House: Bless All the Dear Children at the Internet Movie Database Little House Wiki

It's Garry Shandling's Show

WARNING: Article could not be rendered - ouputting plain text.
Potential causes of the problem are: (a) a bug in the pdf-writer software (b) problematic Mediawiki markup (c) table is too wide

It's Garry Shandling's ShowGenre SitcomCreated byGarry Shandling and Alan ZweibelStarringGarry ShandlingGeoffrey Blake (actor)Geoffrey BlakeMolly CheekJessica HarperScott NemesMichael TucciCo-Starring:Bernadette BirkettIan BuchananBarbara CasonPaul WillsonTheme music composer Joey CarboneCountry of origin United StatesLanguage(s) EnglishNo. of seasons 4No. of episodes 72ProductionExecutive producer(s)Bernie BrillsteinBrad GreyGarry ShandlingProducer(s)Jeff FranklinJim GeoghanAl JeanMike ReissRunning time 30 minutesBroadcastOriginal channelShowtime (TV network)ShowtimeOriginal run September 10, 1986 – May 25, 1990ChronologyFollowed byThe Larry Sanders Show (1992–1998)It's Garry Shandling's Show is an American sitcom which was initially broadcast on Showtime (TV network)Showtime from 1986 in television1986 to 1990 in television1990. It was created by Garry Shandling and Alan Zweibel. The show is notable for its frequent use of breaking the fourth wall to allow characters to speak directly to the audience. Its format inspired Sean Hughes (comedian)Sean Hughes to create Sean's Show in the UK.Synopsis The 30 minute series stars Garry Shandling as, more or less, himself: a Neurosisneurotic, somewhat self-obsessed stand-up comedian, who just happens to be aware he is a TV sitcom character. Garry spends just as much time interacting with the studio audience as he does the regular cast members, offering up opening monologues and show-closing summations of the episode's events (much like George Burns on The Burns and Allen Show). On Garry's show, however, all the supporting characters know they are on a TV show, not just Garry; and the studio audience is often in the storyline. In one episode, Garry invites them to make themselves at home after he leaves the scene (they end up throwing a raucous party); in another, a surprise partysurprise birthday party for Garry's mother goes awry when the audience's shout of "Surprise!" causes her to have a heart attack. In a memorable episode, Garry moves out of town, with his apartment (and show) taken over by Red Buttons; when Garry tries to return, his friends admit they like Red better.At the time of the show's production, Garry Shandling actually lived in Sherman Oaks, CA, just like the character on the TV show. His condominium on the show was styled to be just like his real-life condo, down to the room layout and the furnishings.DVD commentary from It's Garry Shandling's Show, Season 3, Episode 46, "Garry Acts Like A Moron"Storylines were often manipulated by Shandling to create more favorable outcomes, or simply to speed things along (one episode ended years later, for example). On America's presidential election night in 1988, Showtime presented a live episode wherein Shandling brought in Soul Train host Don Cornelius to incorrectly announce that Michael Dukakis had soundly defeated George H. W. Bush.The show's bouncy, well-remembered theme song was "This is the Theme to Garry's Show", sung by Los Angeles musician Bill Lynch.http://mobile.ljworld.com/news/2007/may/04/midwest_icon_bill_lynch_discovers_how_live_music/ The song's lyrics are self-referencing, explaining how the song came to be ("Garry called me up and asked if I would write his theme song") and asking what the listener thinks of it. Considered a critical and niche marketniche success,http://www.museum.tv/archives/etv/I/htmlI/itsgarrysh/itsgarrysh.htm It's Garry Shandling's Show ran 72 episodes and was on the air for four seasons (1986–1990). The show was later picked up by Fox Broadcasting Company from 1988 to 1990 as part of its Sunday night lineup; Fox began airing the show from the beginning, but due to longer seasons for network shows versus cable televisioncable, had caught up by the time the show left Fox in March 1990. New episodes continued for a few months thereafter on Showtime. The show introduced much of the country to Shandling and paved the way for his more successful run as late-night talk show host Larry Sanders on HBO's The Larry Sanders Show.Cast Throughout the series, the cast was divided into the "Starring" cast, whose names were featured in the opening credits; the "Co-Starring" cast and the "Guest Starring" cast, both of which had the actors' names featured in the end credits. StarringGarry Shandling as Garry Shandling - Immature star/host of the show. Constantly obsessed with his hair. Geoffrey Blake (actor)Geoffrey Blake as Lewis (early season 1) - Garry's ladies' man friend. Dropped from the series after the first run of six episodes. Molly Cheek as Nancy Bancroft - Garry's "attractive but nonthreatening platonic neighbor." Jessica Harper as Phoebe Bass (season 4) - Garry's girlfriend and eventual wife. Scott Nemes as Grant Schumacher - Garry's neighbor, teenage son of Pete and Jackie Schumacher. Michael Tucci as Pete Schumacher - Garry's close friend and neighbor. Father of Grant and husband of Jackie. Very nerdy and clumsy. Leads his son's Cub Scout troop. Co-starringBernadette Birkett as Jackie Schumacher (mid-season 1 through season 4, recurring guest early season 1) - Garry's neighbor, Pete's wife and Grant's mother. Ian Buchanan as Ian McFyfer (mid-season 3 through season 4, recurring guest early season 3) - Nancy's boyfriend and eventual husband. Barbara Cason as Ruth Shandling (seasons 2-4, recurring guest season 1) - Garry's elderly mother. Paul Willson as Leonard Smith (mid-season 1 through season 4, recurring guest early season 1) - The owner of Garry's condo complex who frequently drops by uninvited so that he can end up on-camera. He is a nuisance to Garry and his friends. Recurring guest starsBruno Kirby as Brad Brillnick (seasons 3-4) - Garry's agent. Richard Fancy as Mr. Stravely (seasons 3-4) - network boss. Roy Brocksmith as Mr. Guest (seasons 3-4) Chi Ngo as Kim Moon (seasons 3-4) Danny Dayton as Mr. Peck (seasons 2-3) - Comedy club owner. Rob Reiner (seasons 1-3), Tom Petty (seasons 2-3), and Martin Mull (seasons 3-4) as themselves While fighting ovarian cancer, Gilda Radner guest-starred as herself on the show in 1988 in what would be her final television appearance. When Shandling asked her why she had not been seen on television for a while, Radner replied, "Oh, I had cancer. What did you have?"Crew The show was created and written by Garry Shandling and Alan Zweibel. Shandling wrote 15 episodes of the show. Garry, Alan and Garry's manager Brad Grey were the executive producers of the show. Broadcast history In the U.S., the show aired on Showtime (TV network)Showtime: The first season aired on Wednesdays and the rest of the series aired on Sundays. The show was originally picked up for six episodes. After the success of them, Showtime ordered 12 more (although only 10 were produced). Subsequent seasons had differing numbers of episodes. To bolster its Sunday night lineup, FOX secured the rights to air reruns of It's Garry's Shandling's Show not long after they aired on Showtime, so all of America could see what the fuss was about. The reruns aired Sundays on FOX at 9:00-9:30 from March 1988 to July 1989, at 9:30-10:00 in July 1989, at 10:00-10:30 from July to August 1989 and at 10:30-11:00 from August 1989-March 1990.Brooks, Tim (2007). The Complete Directory to Prime Time Network and Cable TV Shows: 1946-Present (Ninth Edition). Ballantine Books. p. 678. ISBN 978-0-345-49773-4.US Season Episodes Premiere Date End Date 1 16 September 10, 1986 April 10, 1987 2 17 October 25, 1987 March 18, 1988 3 20 October 21, 1988 April 21, 1989 4 19 November 10, 1989 May 25, 1990 UK Seasons 1 and 2 of the show were aired in the UK on BBC2 over three series from 1987 to 1990. The airdates did not follow the original episode sequence. An initial season of 10 episodes aired on Sunday nights at 18:50 BST from 5 July-20 September 1987. Two further episodes were aired on 19 November and 3 December, 1987. A second season of 8 episodes, drawn from the end of series 1 and the beginning of series 2 was shown on Tuesday nights at around 22:20 BST from 31 May-2 August 1988. A Christmas special was shown on Christmas Eve 1989, preceding a third and final season, returned to Sunday nights at 18:10 GMT, which ran 18 February-15 April 1990. Season Episodes Premiere Date End Date 1 10 5 July, 1987 20 September, 1987 # 2 19 November, 1987 3 December, 1987 2 8 31 May 1988 2 August 1988 # 1 24 December 1989 3 8 18 February, 1990 15 April, 1990 Awards and nominations The show was nominated for four Primetime Emmy Awards. It won five CableACE Awards (three for Best Comedy Series and two for Garry Shandling). It won an American Comedy AwardsAmerican Comedy Award for Funniest Male Performance in a Comedy Series-Cable or Syndicated, and an award for Outstanding Achievement in Comedy in 1988 from the Television Critics Association. DVD release On October 20, 2009, Shout!

Factory released It's Garry Shandling's Show: The Complete Series on DVD in Region 1. The 16-disc set features extensive bonus features including featurettes, commentaries and outtakes.http://www.tvshowsondvd.com/news/Garry-Shandlings-The-Complete-Series/12333 Time praised the set's release, writing "As self-referential comedies went, you never meta better one","Short List". Time, October 26, 2009, p. 63. while Entertainment Weekly named it one of the best TV on DVD releases of 2009.http://www.ew.com/ew/article/0,,20331989,00.htmlThe first season of the series was released as a stand-alone title in America in April 2010,http://www.tvshowsondvd.com/news/Garry-Shandlings-Season-1/13144 and the first two seasons have been released in the United Kingdom by Fabulous Films (which owns the distribution rights to many of Shout! Factory's releases) with the US sets' extras ported in their entirety. Syndication The show was shown on TV Land Canada until 2010. ReferencesExternal links It's Garry Shandling's Show at the Internet Movie Database It's Garry Shandling's Show at TV.com It's Garry Shandling's Show at Epguidesepguides.com

Comedy_Central

Comedy Central	
Launched	April 1, 1991
Owned by	Viacom
Picture format	480i (SDTV) 1080i (HDTV)
Country	United States
Language	English
Headquarters	345 Hudson Street New York City
Formerly called	CTV: The Comedy Network (April 1 – May 31, 1991)
Replaced	Merger of The Comedy Channel And Ha!
Website	comedycentral.com [1]
Availability	
Satellite	
DirecTV	Channel 249 (SD/HD) Channel 1249 (VOD)
Dish Network	Channel 107 (SD/HD)
Cable	
Available on most cable systems	Check local listings
IPTV	
Verizon FiOS	Channel 190 (SD) Channel 690 (HD)
AT&T U-Verse	Channel 140 (SD) Channel 1140 (HD)

Comedy Central is an American cable television and satellite television channel that carries comedy programming, both original and syndicated.

Since late 2006, Comedy Central has expanded globally with localized channels in Germany, Hungary, India, Israel, Italy, Latin America, New Zealand, The Netherlands, Poland, Sweden, Republic of Ireland, Bosnia and Herzegovina, Croatia, Macedonia, Montenegro, Serbia, Slovenia,[1] Spain, and the United Kingdom. International channels are operated by Viacom International Media Networks.

History

Creation (1989–1991)

On November 15, 1989, Time Warner, owners of HBO, launched The Comedy Channel as the first cable channel devoted exclusively to comedy-based programming. On April 1, 1990, Viacom (who owned MTV, VH1, and Nickelodeon) launched a rival channel called Ha![2] that featured reruns of situation comedies and some original sketch comedy.

The Comedy Channel's programs were originated from the HBO Downtown Studios at 120 East 23rd Street in Manhattan. The format prior to the merger with Ha! included several original and unconventional programs such as *Onion World* with Rich Hall and *Mystery Science Theater 3000*, as well as laid-back variety/talk shows hosted by comedians, including *The Sweet Life* with Rachel Sweet, *Night After Night with Allan Havey*, *Sports Monster* and *The Higgins Boys and Gruber*, the latter of whom performed sketches in between showings of vintage television serials like *Supercar*, *Clutch Cargo*, and *Lancelot Link, Secret Chimp*.

The standard format for The Comedy Channel's shows usually involved the various hosts introducing clips culled from the acts of stand-up comedians as well as classic comedies of the 1970s and 1980s, such as *Young Frankenstein* and *Kentucky Fried Movie*, presented in a style similar to music videos. In the early days, certain hours of the day when clips were shown without "host segments" were dubbed *Short Attention Span Theater*. In 1990, hosts under this title, Jon Stewart and Patty Rosborough, were introduced. Comedian Marc Maron also hosted the series.

While The Comedy Channel broadcast mostly low-budget original programming,[3] Ha!'s schedule featured sitcom and sketch comedy reruns (many of which had been previously licensed for sister network Nick at Nite) as well as complete 90-minute reruns of *Saturday Night Live* from the sixth through sixteenth seasons.

After two years of limited distribution, the two channels merged into one, relaunching on April Fools' Day 1991 as **CTV: The Comedy Network**, changing its name to *Comedy Central* on June 1, 1991 to prevent issues with CTV, a Canadian broadcast television network.[4] Viacom bought out Time Warner's half in April 2003 for $1.23 billion.[5] Despite HBO's exit from the venture, the MTV Networks division in charge of Comedy Central is still called Comedy Partners, currently being a partnership of Viacom International, the subsidiary MTV Networks does business as, and Viacom Hearty Ha! Ha!, the subsidiary that owned Ha! and Viacom's original half of the network.[6]

1991–1997

From the late 1980s through the mid-1990s, much of the programming on Comedy Central, and its predecessors, consisted of comedy films, sitcom reruns, half-hour specials and clip shows featuring comedians. With the exception of the cult favorite *Mystery Science Theater 3000*, the channel had a relatively small viewership. A notable early success was *Politically Incorrect* with Bill Maher, which after showing promise on Comedy Central was quickly snapped up by the ABC network. Additionally, *The Daily Show* had got its start with premiere host Craig Kilborn, although it would take a few more years for the show to reach high popularity (and shift toward a focus on political humor) with the introduction of Jon Stewart (who was former co-host of *Short Attention Span Theater* from 1991).

The original Comedy Central logo used from June 1, 1991 to 7:59 am ET on October 30, 2000. An earlier variant of this logo has the "Comedy Central" text bigger, almost taking up the marquee sign. That variant lasted until 1995.

Dr. Katz, Professional Therapist was also a notable original program from this era, as well as the game show *Win Ben Stein's Money*. Successful non-original programming included Canadian comedy group *The Kids in the Hall* and British shows such as the U.K. edition of *Whose Line Is It Anyway?* (predecessor of the U.S. version, featuring much of the same American cast as would later be seen in the U.S.) and the sitcom *Absolutely Fabulous*. Some later seasons of *"AbFab"*, as it was informally known, were partially financed by Comedy Central. Comedy Central also aired nationwide reruns of Seattle's *Almost Live!* between 1992 and 1993.

South Park

The channel made a breakthrough when *South Park* premiered in 1997. Being the first major cable show to carry the TV-MA rating for mature audiences, the show was too controversial to be picked up by a mainstream network.[7] As word of mouth spread, the number of people who requested that Comedy Central be added to their cable systems increased, and the channel became available in over 50% of American homes by 1998.

South Park's main characters: (L-R) Eric Cartman, Kyle Broflowski, Stan Marsh and Kenny McCormick

2000–2003

The network's second logo used from October 30, 2000 until New Years Eve 2010. Used in the United Kingdom until July 31, 2012.

At 8 am ET on October 30, 2000, Comedy Central modernized its globe logo, by straightening the buildings, modernizing the globe, and removing the transmitter from the logo. The management of the network said that the transmitter of the 1991 logo was said to "communicate the 1950s broadcast era". In 2002, Comedy Central Records was formed as a means of releasing albums by comedians that have appeared on the network.[8] This unit would ultimately be sold to Fox.

Since 2003, Comedy Central has created a tradition of roasting comedians in tradition of the New York Friars' Club roasts. During these roasts friends of the roastee, along with comedians, take turns making fun of the roastee, the other roasters and occasionally audience members. So far, the roastees have included Denis Leary,[9] Jeff Foxworthy,[10] Pamela Anderson,[11] William Shatner,[12] Flavor Flav,[13] Bob Saget,[14] Larry the Cable Guy,[15] Joan Rivers, David Hasselhoff,[16] Donald Trump, Charlie Sheen, and Roseanne Barr.

The *Secret Stash*

The success of *South Park*, despite its mature content, encouraged the network to continue to push the limits on adult language. Every Saturday and Sunday morning at 1 am eastern time/12 am central time, a movie, comedy special or animated program is shown unedited for language. This is called the ***Secret Stash***. It premiered on July 4, 2003 (with the unedited cable television debut of *South Park: Bigger, Longer & Uncut*). Though no language is censored on the Secret Stash, most nudity in the programs is still edited out, with the exception of limited nudity allowed in animated programs (such as *Drawn Together*), Rear Nudity and the occasional topless scene.

2004–2006

In late 2004, it was reported that the top four rated shows on Comedy Central were, in ascending order, *South Park*, *Chappelle's Show*, *The Daily Show*, and *Reno 911!*. Shortly thereafter, Dave Chappelle backed out of the much-anticipated third season of *Chappelle's Show*.[17] Meanwhile, *The Daily Show* continued to climb in ratings. In October 2005, on the occasion of a new three-year contract for *South Park* and the launch of *Daily Show* spin-off *The Colbert Report*, it was reported that *South Park* and *The Daily Show* were the two highest rated shows on Comedy Central. Comedy Central chief Doug Herzog was reported as saying that he hoped to continue to air new seasons of *South Park* forever, and that *The Colbert Report* fulfilled a long-held plan to extend the *Daily Show* brand.

On April 5, 2006, in a controversial two-part episode arc titled "Cartoon Wars Part I" and "Cartoon Wars Part II", *South Park* touched the issue of the recent protest over the Danish cartoon drawings depicting the Muslim prophet Muhammad. The image of Muhammad did not appear in the episode. On April 13, 2006, Comedy Central issued a statement[18] which appears to confirm that the network prohibited the show's creators from airing an image of Muhammad. The statement reads, *"In light of recent world events, we feel we made the right decision."* An anonymous source close to the show indicated that *South Park* creators Trey Parker and Matt Stone were informed of the policy several weeks earlier, and wrote this story arc in protest. This is a change of policy for Comedy Central, having allowed *South Park* to portray an image of Muhammad in an earlier episode, "Super Best Friends". Oddly enough, an image of Muhammad was still briefly visible in the opening credits of the "Cartoon Wars" episodes (the image had been there as a call-back to "Super Best Friends").

2007–2010

On Monday January 15, 2007, MTV Networks International launched Comedy Central in Germany which is available for free throughout Europe. The channel airs 33 shows either dubbed in German or subtitled while also airing locally produced shows.[19] As of April 30, Dutch channel The Box has been transformed into the Dutch version of Comedy Central during the primetime and overnight hours timesharing with Nickelodeon.[20] And on May 1, 2007 the channel has launched in Italy replacing Paramount Comedy.[21]

As of June 27, 2007, CTVglobemedia-owned networks CTV and The Comedy Network have exclusive Canadian rights to the entire Comedy Central library of past and current programs on all electronic platforms, under a multi-year agreement with Viacom, expanding on past programming agreements between the two channels. Canadian users attempting to visit Comedy Central websites will also be redirected to The Comedy Network's website. The Canadian channel will keep its own brand name, but the agreement is otherwise very similar to the earlier CTV/Viacom deal for MTV in Canada.[22] As of 2011 this Geocaching no longer applies and both the Comedy Central and The Comedy Network website can be access worldwide with the exception of videos which remain only accessible within eaches native country.

In December 2007, Comedy Central picked up a show hosted by Lewis Black called *Lewis Black's Root of All Evil.*[23] The show didn't start playing on Comedy Central until March 2008. On January 9, 2008, it was announced the Comedy Central and MTV would put programs online for free starting in February.[24] On January 24, Scott Landsman became the Vice President of Original Programming and Development at the network.[25]

On March 27, 2008, the Swedish Radio and TV Authority approved an application from Comedy Central regarding being allowed to air television programs in Sweden. The grant allows Comedy Central to broadcast on the terrestrial television network between January 1, 2009, and March 31, 2014, after which a new request must be submitted in order to continue broadcasting.[26] The main Comedy Central network in the United States picked up a remake of *The Gong Show* hosted by Dave Attell,[27] star of his former self-titled Comedy Central series *Insomniac* for premiere in July 2008. Another new show called *Reality Bites Back*[28] premiered after *The Gong Show with Dave Attell.*

In June 2008, Comedy Central picked up sketch comedy show *Important Things with Demetri Martin* which began airing in February 2009.[29] On April 1, 2009, Comedy Central began airing in New Zealand as Channel 010 on SKY Digital. On April 6, Paramount Comedy in the UK and Ireland rebranded as Comedy Central. On April 7, 2009, it was announced Comedy Central would air new stand-up comedy specials starring Christopher Titus, Gabriel Iglesias, Pablo Francisco, Jim Breuer, Mitch Fatel and Pete Correale, and ventriloquist Jeff Dunham.[30] An animated show entitled *Ugly Americans* has also been picked up by the network.[31] In 2009, *The Goode Family*[32] premiered. Also in 2009, Thomas Lennon announced via Twitter that *Reno 911!* had been cancelled[33] after six successful seasons, much to outraged fan disapproval. The network has also played a role in the revival of the animated series *Futurama*, which Fox had cancelled in 2003. New episodes began airing on Comedy Central in 2010.[34]

South Park episodes "200" and "201" aired in April 2010, revisiting the issue of the Islamic religious figure Muhammad's perceived immunity to parody, for fear of violent retaliation. The Super Best Friends return, but Muhammad is entirely covered by a black bar reading "CENSORED" through all of his screen time. By the second episode of the two-parter, Comedy Central decided to censor every instance of his name, as well as three entire monologues, from the end of the show. The monologues dealt with the subjects of censorship and intimidation, but did not actually use Muhammad's name. Parker and Stone have since issued a statement to the press, confirming that the "bleeps" were added weeks after the show was finished, and that Comedy Central has refused to let them post the original version to South Park Studios, in addition to retroactively removing the original "Super Best Friends" episode.[35]

2011–present

On December 10, 2010, Comedy Central introduced a new logo for the network that launched on January 1, 2011, which left behind the previous theme of a world-sized "tower" broadcasting the network/skyscrapers, in favor of an image of two "C"'s, with one of them and the word "Central" turned upside-down within the new logomark. The new logo was designed to represent the network's unique brand of comedy (with some drawing comparisons to the copyright symbol as inspiration for its design and use), and to provide the network with a logo that could be easily used across different platforms, such as social media.[36] [37] This new logo has won several industry awards.[38]

Poland was the first who started the international rebrand in Europe on February 20, 2011; and then Hungary on April 1, 2011. The next two countries were the Netherlands and Germany on October 1, 2011. Comedy Central New Zealand rebranded in April 2012.

Viacom 18 launched the channel in India on 23 January 2012.[39]

StarHub launched Comedy Central Asia in Singapore on 1 November 2012; the channel was added to its Basic Entertainment Upsize group.[40]

Comedy Central HD

Comedy Central HD is a 1080i high definition simulcast of Comedy Central that launched on January 13, 2009.[41] Cablevision was the first carrier to add it. DirecTV added the HD version on January 21, 2009. Dish Network added the HD version on April 9. 2009.[42] It was also reported that some Cox Cable systems would get it by the end of January 2009.[41] Comedy Central anticipated that several more providers would add it later in that year.[41]

Logo of Comedy Central HD

Currently, original shows that air in HD include *Tosh.0*, and newer episodes of *South Park*, *Comedy Central Presents*, *Reno 911!*, as well as *Workaholics* and *The Jeff Dunham Show*. The new season of *Key & Peele* and other stand up specials, as well as movies, air in HD as well.[41]

The Daily Show and *The Colbert Report* began airing in HD on January 4, 2010.[43] Many episodes of *South Park* that were originally made in 4:3 standard definition were re-rendered to air in a 16:9 format on Comedy Central HD. Comedy Central HD currently only airs in Eastern Standard Time.

International

Local versions of *Comedy Central* include:

- Germany
- India
- Israel
- Italy
- Latin America
- New Zealand
- The Netherlands
- Poland
- Sweden
- Hungary
- United Kingdom / Ireland
- Africa[44]
- Croatia
- Asia

Comedy Central Family:

- The Netherlands
- Poland

Comedy Central Extra:

- The Netherlands
- Bosnia and Herzegovina, Croatia, Macedonia, Montenegro, Serbia and Slovenia[1]
- United Kingdom and Ireland

Criticism

Comedy Central has been a frequent target of criticism from the conservative group Parents Television Council, which accuses them of bigotry and blasphemy,[45] [46] especially within the programs *South Park*, *The Sarah Silverman Program*, *Halfway Home*, and the annual "Roast" special.[47] PTC has used their criticisms against Comedy Central for their support of the Family and Consumer Choice Act of 2007, which would allow American cable TV subscribers to choose which channels they subscribe to,[48] and to persuade advertisers to stop advertising on the channel.[49] PTC founder and former president L. Brent Bozell III has called the channel unfunny, claiming the channel has managed "to reach the top of its field in spite of – or, better put, because of – the network's sheer lack of comedic talent" by its "extensive reliance on shocking or disgusting humor".[50] The PTC attacked the channel[51] for airing advertisements for "Girls Gone Wild". The channel airs the least censored version of the film *Not Another Teen Movie*, as well as uncut versions of films such as *Coming to America*, *Dogma* and Jay and Silent Bob Strike Back.

On November 5, 2007, an open letter[52] was written by VideoSift to protest publicly the blocking of Comedy Central's embedded video content for non-U.S.-based viewers.

On April 21, 2010, Comedy Central censored the *South Park* episode, "201", in response to a death threat issued by users of a radical Muslim website over the episode's planned depiction of the Islamic prophet Muhammad, which led several newspaper columnists to condemn the network's actions as tantamount to abetting terrorism. As a result, "201" and the episode that preceded it were heavily edited and not shown in repeats.

References

[1] "Viacom to Launch Comedy Central Extra in Adriatic Region" (http://www.hollywoodreporter.com/news/viacom-comedy-central-extra-launch-adriatic-357169). The Hollywood Reporter. 2012-07-31. . Retrieved 2012-10-27.

[2] latimes.com (http://articles.latimes.com/1989-11-15/entertainment/ca-1886_1_comedy-channel) Cable Comedy—Will HBO Have the Last Laugh? Television: The 24-hour Comedy Channel premieres tonight, but Viacom has plans to launch its own comedy channel, HA!, in the spring.

[3] Boone, Brian (January 12, 2012). "The Origin and Early Programs of Comedy Central" (http://splitsider.com/2012/01/the-origin-and-early-programs-of-comedy-central). *Splitsider*. . Retrieved March 29, 2012.

[4] (Name change to Comedy Central within a month of launch due to a lawsuit with CTV in Canada) (http://findarticles.com/p/articles/mi_hb4895/is_199103/ai_n17999383)

[5] "Viacom buys Comedy Central" (http://money.cnn.com/2003/04/22/news/companies/viacom/index.htm). Reuters. April 22, 2003. . Retrieved 2007-08-03.

[6] http://tess2.uspto.gov/bin/showfield?f=doc&state=4002:b66558.2.1

[7] Carter, Bill. MEDIA: BROADCASTING; Comedy Central makes the most of an irreverent, and profitable, new cartoon hit (http://query.nytimes.com/gst/fullpage.html?res=9C0DE5D61239F933A25752C1A961958260&sec=&spon=&pagewanted=2) (Part 2 of 2). *The New York Times*: Nov 10, 1997.

[8] comedycentral.com (http://www.comedycentral.com/press/series/comedycentralrecords.jhtml) Press Central – Comedy Central Records

[9] "Official site" (http://www.comedycentral.com/shows/leary_roast/index.jhtml). Comedy Central. . Retrieved November 13, 2011.

[10] "Comedy Central Roast of Jeff Foxworthy" (http://www.comedycentral.com/shows/roast_foxworthy/index.jhtml). Comedy Central. . Retrieved November 13, 2011.

[11] "Comedy Central Roast of Pamela Anderson" (http://www.comedycentral.com/shows/roast_anderson/index.jhtml). Comedy Central. . Retrieved November 13, 2011.

[12] "Roast of Shatner" (http://www.comedycentral.com/shows/roast_shatner/index.jhtml). Comedy Central. . Retrieved November 13, 2011.

[13] "Comedy Central Roast of Flavor Flav official site" (http://www.comedycentral.com/shows/roast_flavor_flav/index.jhtml). Comedy Central. . Retrieved November 13, 2011.

[14] "Official site" (http://www.comedycentral.com/shows/roast_saget/index.jhtml). Comedy Central. . Retrieved November 13, 2011.

[15] "Official site" (http://www.comedycentral.com/shows/roast_larry/index.jhtml). Comedy Central. . Retrieved November 13, 2011.

[16] "Official site" (http://www.comedycentral.com/shows/roast-david-hasselhoff/index.jhtml). Comedy Central. . Retrieved November 13, 2011.

[17] Comedy Central to air Chappelle remainders (http://www.msnbc.msn.com/id/10337064/) – MSNBC – Dec 12, 2005

[18] "'South Park' Creators Skewer Own Network" (http://www.newsvine.com/_news/2006/04/13/165631-south-park-creators-skewer-own-network). Newsvine. April 13, 2006. . Retrieved November 13, 2011.

[19] Tzortzis, Andreas (February 18, 2007). "Comedy Central plays to a German sense of humor" (http://www.iht.com/articles/2007/02/18/business/comedy.php). *International Herald Tribune*. . Retrieved 2007-08-03.

[20] 16.06 Europe/London, March 23, 2007 (March 23, 2007). "Dutch launch for Comedy Central" (http://www.broadbandtvnews.com/today/?p=573). Broadbandtvnews.com. . Retrieved November 13, 2011.

[21] http://www.broadcastclub.co.uk/sateuropa_forum/upload/viewtopic.php?pid=318

[22] CTV Strikes Multi-Platform Content Deal With Comedy Central (http://www.cnw.ca/en/releases/archive/June2007/27/c9701.html), I Have A Very Large Penis CTV press release, June 27, 2007

[23] tvsquad.com (http://www.tvsquad.com/2007/10/05/comedy-central-picks-up-lewis-blacks-show/) Comedy Central picks up Lewis Black's show

[24] Jones, K.C. (January 9, 2008). "Free MTV And Comedy Central Online" (http://www.informationweek.com/news/showArticle.jhtml?articleID=205601394). Information Week. . Retrieved 2008-01-09.

[25] Eggerton, John (January 24, 2007). "Landsman Gets VP Stripes at Comedy Central" (http://www.broadcastingcable.com/article/CA6525373.html?rssid=193). *Broadcasting & Cable*. . Retrieved 2007-01-27.

[26] "Announcement regarding new DVB-T channels going live in Sweden on April 1, 2008" (http://www.rtvv.se/se/Press/Nyheter/080327/). . Retrieved 2008-03-27.

[27] zap2it.com (http://www.zap2it.com/tv/news/zap-comedycentralgongshowdaveattell,0,4414601.story) Comedy Central Resurrects 'The Gong Show'

[28] multichannel.com (http://www.multichannel.com/article/84272-Comedians_Square_Off_In_Reality_Bites_Back_Series.php) Comedians Square Off In 'Reality Bites Back' Series – Comedy Central's First Unscripted Competition Series Mocks Reality Genre

[29] "Stewart stamp on 'Martin'" (http://www.hollywoodreporter.com/hr/content_display/news/e3iff0ec139962d42d4152b670707cce783). Hollywoodreporter.com. October 3, 2007. . Retrieved 2009-02-16.

[30] Lafayette, Jon. "TV Week April 7, 2009 Comedy Central Commits to Stand-Up Specials" (http://www.tvweek.com/news/2009/04/comedy_central_commits_to_stan.php). Tvweek.com. . Retrieved November 13, 2011.

[31] "Comedy Central Orders Animated Show, Gets Righteous" (http://www.thrfeed.com/2009/05/comedy-central-ugly-americans-.html#more). The Live Feed. May 14, 2009. .

[32] latimes.com (http://articles.latimes.com/2010/jan/04/entertainment/la-et-goode-family4-2010jan04) New life for 'Goode Family' – Canceled last season by ABC, the series is getting another chance to catch on, this time via Comedy Central.
[33] gawker.com (http://gawker.com/5337237/reno-911-cancelled-by-comedy-central) Reno 911! Cancelled By Comedy Central
[34] movieweb.com (http://www.movieweb.com/news/NEqF9susTJPtut\) Futurama Finally Returns with Brand New Episodes in June!
[35] "201 (South Park) – Wikipedia, the free encyclopedia" (http://en.wikipedia.org/wiki/201_(South_Park)#cite_note-NYT_Arts_Blog-2). En.wikipedia.org. . Retrieved November 13, 2011.
[36] "Comedy Central Unveils Serious New Logo – Vulture" (http://nymag.com/daily/entertainment/2010/12/comedy_central_unveils_serious.html). Nymag.com. . Retrieved November 13, 2011.
[37] "Comedy Central Flips With New Logo – 2010-12-09 18:55:26 | Broadcasting & Cable" (http://www.broadcastingcable.com/article/460888-Comedy_Central_Flips_With_New_Logo.php). Broadcastingcable.com. . Retrieved November 13, 2011.
[38] "Thelab : thelog :" (http://www.thelabnyc.com/blog/article/thelab-wins-awards). Thelabnyc.com. February 7, 2011. . Retrieved November 13, 2011.
[39] Pereira, Priyanka (February 17, 2012). "For a Few Laughs" (http://www.indianexpress.com/news/for-a-few-laughs/913146/). *The Indian Express*. . Retrieved March 10, 2012.
[40] "Comedy Central Asia, Asia's First and Only 24/7 All-Comedy Network Available to All StarHub TV Subscribers" (http://www.starhub.com/content/corporate/newsroom/2012/10/comedy-central-asia--asia-s-first-and-only-24-7-all-comedy-netwo.html). .
[41] "Multichannel News January 13, 2009 Comedy Central Launches HDTV Network – New Service Available on Cablevision Systems with Cox, DirecTV to Come" (http://www.multichannel.com/article/CA6629076.html?nid=2226&rid=reg_visitor_id&source=link). Multichannel.com. . Retrieved November 13, 2011.
[42] Gelman, Vlada. "TV Week April 10, 2009 Dish Network Adds BET, MTV Channels in HD" (http://www.tvweek.com/news/2009/04/dish_adds_bet_mtv_channels_in.php). Tvweek.com. . Retrieved November 13, 2011.
[43] "Multichannel News December 17, 2009 'Daily Show,' 'Colbert Report' To Ring In 2010 In HD – Comedy Central Series Will Be Delivered In HD, Beginning Jan. 4" (http://www.multichannel.com/article/440934-_Daily_Show_Colbert_Report_To_Ring_In_2010_In_HD.php). Multichannel.com. . Retrieved November 13, 2011.
[44] http://www.comedycentralafrica.com/Comedy Central Africa website
[45] Bozell, L. Brent III (December 22, 2005). "Comedy Central's War on Christmas" (http://web.archive.org/web/20080201043432/http://www.cnsnews.com/bozellcolumn/archive/2005/col20051222.asp). *CNS News*. Creators Syndicate. Archived from the original (http://www.cnsnews.com/bozellcolumn/archive/2005/col20051222.asp) on February 1, 2008. . Retrieved 2007-08-01.
[46] "Comedy Central Shows God in One-Night Stand" (http://www.parentstv.org/PTC/publications/release/2007/0309.asp) (Press release). Parents Television Council. March 9, 2007. . Retrieved 2007-08-01.
[47] Bozell, L. Brent III (August 24, 2006). "Roasting the Final Frontier" (http://www.mrc.org/BozellColumns/entertainmentcolumn/2006/col20060824.asp). *MRC.org*. Creators Syndicate. . Retrieved 2007-08-01.
[48] Isett, Dan (June 14, 2007). "Remarks Presented by Dan Isett of the PTC at the News Conference Regarding the "Family and Consumer Choice Act of 2007"" (http://web.archive.org/web/20070930200903/http://www.parentstv.org/ptc/publications/news/cablebill-danisett.asp). Parents Television Council. Archived from the original (http://www.parentstv.org/ptc/publications/news/cablebill-danisett.asp) on September 30, 2007. . Retrieved 2007-08-01.
[49] Bozell, L. Brent III (September 6, 2005). "Letter to Advertisers Concerning their Sponsorship of the August 16th Roast of Pamela Anderson on Comedy Central" (http://www.parentstv.org/PTC/advertisers/letters/Various_Roast_090605.htm). Parents Television Council. . Retrieved 2007-08-01.
[50] Bozell, L. Brent III (March 24, 2006). "The Arrested Adolescent's Channel" (http://web.archive.org/web/20070611135632/http://www.cnsnews.com/bozellcolumn/archive/2006/col20060324.asp). *CNS News*. Creators Syndicate. Archived from the original (http://www.cnsnews.com/bozellcolumn/archive/2006/col20060324.asp) on June 11, 2007. . Retrieved 2007-08-01.
[51] Parents Television Council – Letters to the Editor – Offensive Ads (http://www.parentstv.org/ptc/letters/main.asp#4)
[52] "VideoSifts open letter to Comedy Central and Viacom" (http://www.videosift.com/talk/An-Open-Letter-to-Comedy-Central-and-Viacom). Videosift.com. . Retrieved November 13, 2011.

External links

- ComedyCentral.com (http://www.comedycentral.com)
- Comedy Central (https://plus.google.com/+ComedyCentral/about) on Google+
- ComedyCentral Latin America (http://www.comedycentral.la)
- ComedyCentral Germany (http://www.comedycentral.de)
- ComedyCentral India (http://www.comedycentralindia.com)
- ComedyCentral Italia (http://www.comedycentral.it)
- ComedyCentral Netherlands (http://www.comedycentral.nl)
- ComedyCentral Poland (http://www.comedycentral.pl)
- ComedyCentral Hungary (http://www.comedycentral.hu)

- ComedyCentral Sweden (http://www.comedycentral.tv)
- ComedyCentral Ireland (http://www.comedycentral.ie)
- Comedy Central New Zealand (http://www.comedycentral.co.nz)
- Comedy Central UK (http://www.comedycentral.co.uk)
- Paramount Comedy Spain (http://www.paramountcomedy.es)

Knots Landing

WARNING: Article could not be rendered - ouputting plain text.

Potential causes of the problem are: (a) a bug in the pdf-writer software (b) problematic Mediawiki markup (c) table is too wide

Knots LandingKnots Landing Logo (Seasons 9–10)Format Soap operaCreated byDavid Jacobs (writer)David JacobsStarringTeri AustinAlec BaldwinTonya CroweWilliam DevaneKevin DobsonStacy GalinaJulie HarrisLisa HartmanJames HoughtonKim LankfordMichele LeeClaudia LonowConstance McCashinDonna MillsDon Murray (actor)Don MurrayKathleen NoonePatrick PetersenMichelle PhillipsJohn PleshetteLarry Riley (actor)Larry RileyTed ShackelfordDouglas SheehanNicollette SheridanJoan Van ArkCountry of origin USANo. of seasons 14No. of episodes 344 (List of Knots Landing episodesList of episodes)ProductionRunning time 47 to 50 MinutesProduction company(s) Roundelay-MF ProductionsLorimar ProductionsDistributor Warner Bros. Television (DVD and Syndication)BroadcastOriginal channelCBS NetworkCBSOriginal run December 27, 1979 – May 13, 1993ChronologyFollowed byKnots Landing: Back to the Cul-de-Sac (1997)Related showsDallas (1978 TV series)Dallas Knots Landing is an American primetime television soap opera that aired from December 27, 1979 to May 13, 1993 on CBS and was a spin-off of Dallas (1978 TV series)Dallas. Set in a fictitious coastal suburb of Los Angeles in California, the show centered on the lives of four married couples living in a cul-de-sac, Seaview Circle. Initially intended to be a Scenes from a Marriage-type drama series, storylines also included rape, murder, kidnapping, assassinations, drug smuggling, corporate intrigue and criminal investigations. By the time of its conclusion, Knots Landing had become one of the longest-running primetime dramas on U.S. television after Gunsmoke, Law & Order, and tied for third place with Dallas and Bonanza (although Bonanza had made far more episodes).Both Bonanza and Knots Landing were eclipsed in 2004 by Law & Order.Knots Landing was created by David Jacobs (writer)David Jacobs (one-time writer of Family (TV series)Family and later producer of Lois & Clark: The New Adventures of Superman) in conjunction with producer Michael Filerman (who would also later co-produce Falcon Crest). Although it became a spin-off of Dallas (1978 TV series)Dallas, the concept predates that series, and was initially rebuffed by CBS in 1977, as the network wanted something more "saga-like." Jacobs then created Dallas, which the network accepted and premiered in 1978. After Dallas became a hit, Jacobs was then able to adapt Knots Landing as a spin-off series by way of incorporating characters originally introduced in the parent series. The series was loosely inspired by the 1973 Ingmar Bergman television miniseries Scenes from a Marriage, but with four marriages. Though initially not as popular as Dallas, Knots Landing eventually outlasted it and garnered much critical acclaim. The series peaked during the 1983–84 season with a 20.8 rating (finishing in 11th place) and a 20.0 rating for the 1984–85 season (when it finished 9th). This can be attributed, in part, to more dramatic storylines as the series became more soap-opera like, and the gradual inclusion of newer characters to interact with the original cast. By the 1988–89 season, Knots Landing was ahead of Dallas in the ratings, though audiences for both shows by this time were less than their earlier years. In 1997, much of the cast reunited for a two-part mini-series entitled Knots Landing: Back to the Cul-de-Sac. In 2005, they reunited again for the non-fiction special Knots Landing Reunion: Together Again in which the cast reminisced about their time on the show. Dallas itself was Dallas (2012 TV series)revived in 2012, with several characters from Knots Landing scheduled to appear in its second season.BackgroundGarrison "Gary" EwingGary Ewing was the middle son and the black sheep (term)black sheep of the Ewing family from Dallas. Gary was an alcoholic, whose father John Ross "Jock" Ewing, Sr.Jock and older brother J. R. EwingJ.R. had never treated as an equal. The insecure Gary met Valene "Val" Clements EwingValene when they were aged 17 and 15 years old respectively. They were married briefly, and Gary left Southfork Ranch and later divorced Valene. With Gary gone from Southfork Ranch, J.R. had Valene followed and 'run out of town' as he took her daughter and manipulated Gary away from her. Years later, Valene and her daughter Lucy Ewing CooperLucy reconnected, causing Valene and Gary to reunite. They remarried and Gary's mother, Miss Ellie EwingMiss Ellie, bought the couple a house in California. Knots Landing is officially spun-off from Dallas in the third season episode entitled "Return Engagement."Credited castMichele Lee (Original cast) as Karen Cooper Fairgate MacKenzie First married to Sid Fairgate, and later to Mack McKenzie, and the mother of Diana, Eric and Michael Fairgate. She and Mack later adopted Meg, (Laura's daughter with Greg Sumner). Karen worked as a community activist and later proved herself to be a good businesswoman by running Knots Landing Motors and Lotus Point, of which she was a joint-owner with Gary and Abby. (Michele Lee would also become the only cast member to appear in all 344 episodes which was a record for most appearances of a female character on American primetime television. The record was surpassed in 2008 by S. Epatha Merkerson's character on NBC's Law & Order).Don Murray (actor)Don Murray (Original cast) as William Sidney Fairgate (1979–1981).Karen's first husband, and the owner of Knots Landing Motors. Hard working, fair and at times stubborn, Sid was the emotional rock of the show during the first two seasons. Sid was killed when his car was sabotaged. Ted Shackelford (Original cast) as Gary EwingGarrison Arthur "Gary" Ewing.The son of John Ross "Jock" Ewing, Sr.Jock and Miss Ellie EwingEllie Ewing, first introduced in the TV series Dallas. Gary, the middle Ewing son between J.R. EwingJ.R. and Bobby James EwingBobby, was the family blacksheep and a recovering alcoholic. The true love of Valene's life and father of their daughter Lucy Ewing CooperLucy (from Dallas) and twins Bobby & Betsy Ewing. After his arrival in California, he began working for Sid Fairgate at Knots Landing Motors, where he was eventually promoted to vice president. He was later fired by Karen after Sid's death, and his marriage to Val was wrecked by his affair with Abby. After his father (Jock Ewing) died, he inherited $10 million and later married Abby, though the marriage ended after a couple of years. After a relationship with the scheming Jill Bennett, he later remarried Valene. (Ted Shackleford is the only castmember along with Michele Lee to remain in the series throughout its entire run, but he did not appear in every episode)Joan Van Ark (Original cast) as Valene EwingValene "Val" Clements Ewing Ewing Gibson Waleska Ewing (1979–1992, 1993).The true love of Gary's life, she is the mother of his three children, including Dallas' Lucy Ewing (Charlene Tilton) and twins Bobby & Betsy Ewing. Valene originally came from Tennessee, and married Gary while they were still in their teens, but she was outcast from the Ewing family by J.R. EwingJ.R.. After she and Gary remarried, they moved to Knots Landing where she became Karen's neighbor and best friend. After their marriage failed again, she later married journalist Ben Gibson, and then psychotic Danny Waleska, before finally marrying Gary for the third time. (Joan Van Ark appears in all but six of the episodes for her 13 seasons, and returned for the series two-part finale in 1993)Constance McCashin (Original cast) as Laura Avery Sumner (1979–1987).The wife of Richard Avery, and later, Greg Sumner. Initially an oppressed housewife, she later became a successful real estate agent. Laura died in 1987 of a brain tumor. John Pleshette (Original cast) as Richard Avery (1979–1983, 1987).A lawyer, and later a restaurant owner. Unhappily married to Laura, he struggled to assert himself after she launched a successful career in real-estate. In 1982, he suffered a nervous breakdown (where he held Laura hostage at gunpoint) and left town in 1983 without his family, as he and Laura divorced. Kim Lankford (Original cast) as Ginger Ward (1979–1983).A kindergarten teacher who wanted to be a singer, she was jealous of her husband's involvement in Ciji Dunne's career. James Houghton (Original cast) as Kenny Ward (1979–1983).A young record producer who, at first, cheated on his wife Ginger, but reformed upon the birth of their daughter Erin Molly. Both he and Ginger moved away from Knots Landing in 1983. Patrick Petersen (Original cast) as Michael Fairgate (1979–1991).Sid and Karen's younger son who would later have an affair with his brother Eric's wife, Linda. Claudia Lonow (Original cast) as Diana Fairgate (1979–1984, 1993).Karen's daughter who fell in love with the evil Chip Roberts (Michael Sabatino). She eventually left Knots Landing to study in New York City. Julie Harris as Lilimae Clements (1980, 1981–1987). Valene's mother. Though they'd been estranged for many years, after Lilimae virtually abandoned Valene as a teenager in her quest to become a country singer, Lilimae reappeared in Valene's life during Season 1. In Season 3, she moved in with Valene and Gary, remaining in Knots Landing until Season 7. Donna Mills as Abby Fairgate Cunningham Ewing Sumner (1980–1989, 1993). Sid's sister, and the neighborhood troublemaker. After divorcing her first husband, Jeff Cunnningham, Abby decided to move to Knots Landing where she began working as a bookkeeper at Knots Landing Motors. She engaged in affairs with Richard, and later Gary, destroying his marriage to Val. Abby then married Gary herself (mainly to get his inheritance) but they divorced a couple of years later, though by this time she was working with both Gary and her former sister-in-law, Karen, as joint-owners of Lotus Point. She later married Greg Sumner (for political reasons) but the marriage lasted only a few months before Abby moved to Japan. Although Abby adored her brother Sid, Abby and Karen rarely got along. Donna Mills appeared in almost all the episodes of the series for her 9 seasons, with the exception of one episode during the 1988–89 season. Tonya Crowe as Olivia Cunningham Dyer (1980–1990). Abby's rebellious daughter who became addicted to drugs. She later married mobster's nephew Harold Dyer against her mother's wishes, making Abby cut her off financially. Kevin Dobson as Marion Patrick "Mack" MacKenzie (1982–1993). Karen's second husband; an attorney who worked for the Governor's office before opening his own private practice. He first met Karen when he worked as the Federal Prosecutor, helping to catch the men responsible for Sid's death. He later worked for his old classmate, Senator Greg Sumner, at his crime commission though later quit because he could not tolerate Sumner's schemes which led to the pair of them becoming rivals. (From joining the series in Season 4, Kevin Dobson appears in all the episodes of the series with the exception of two, during the final season). Lisa Hartman as Ciji Dunne (1982–1983)/Cathy Geary Rush (1983–1986). Ciji and Cathy were both singers. Ciji was murdered, leaving almost everyone in Knots Landing as a suspect. Some months later, Ciji's doppelgänger Cathy Geary showed up after being released from prison. William Devane as Greg Sumner (1983–1993). Mack's ex-best friend and classmate who attended law school with him and even offered him a job working for the crime commission. Initially a politician, Greg became a businessman after inheriting his father's corporation. He often plotted with Abby, whom he later married. Douglas Sheehan as Ben Gibson (1983–1987). Valene's second husband. He was a journalist, who later worked at Abby's television station but disappeared in South America. Alec Baldwin as Joshua Rush (1984–1985). Lilimae's son whom she abandoned as a baby. He was raised by his preacher father and came to Knots Landing as a rather innocent young preacher. He later became a televangelist and married Cathy Geary, but became mentally ill and extremely violent. After he tried to kill Cathy, he accidentally fell off a roof and died. Teri Austin as Jill Bennett (British actress)Jill Bennett (1985–1989). An ex-colleague of Mack's and later Gary's mistress who later tried to murder Valene. She accidentally killed herself after locking herself in the trunk of Gary's car so it would seem like he'd kidnapped her. Nicollette Sheridan as Paige Matheson. (1986–1993). Mack's daughter with Anne Matheson. She became involved with Greg Sumner (whom she also worked for), police detective Tom Ryan, and also Pierce Lawton who tried to kill her. Michelle Phillips as Anne Winston Matheson Sumner (1987, 1989, 1990–1993). Paige's mother and Mack's first love. She initially tried to split up Mack and Karen, but after failing she turned her attentions to Greg and also attempted to cheat Paige out of her inheritance from her grandfather. Larry Riley (actor)Larry Riley as Frank Williams (1988–1992) Frank moved to Knots Landing with his wife Pat and daughter Julie as part of the witness protection program. He eventually took a job in Mack's law firm. Stacy Galina as Kate Whittaker (1990–1993) Claudia's daughter who resembled Greg's late daughter Mary Frances. She was a semi-professional tennis player, but broke her arm which ended her budding career. She later became romantically involved with Gary Ewing. Kathleen Noone as Claudia Sumner Whittaker (1990–1993) Greg Sumner's sister, who moved to town under the guise of accompanying her daughter for college; in reality, she wanted to meddle in Greg's affairs. Other recurring charactersSteve Shaw (actor)Steve Shaw (Original cast) as Eric Fairgate (1979–1990)Karen and Sid's eldest son. Robert JayneBobby Jacoby as Brian Cunningham #1 (1980–1985) Stephen Macht as Joe Cooper (Karen's brother) (1981–1982) Michael Sabatino as Chip Roberts (aka Tony Fenice) (1982–1983) Diana's husband and the murderer of Ciji. He died when he accidentally fell onto a pitchfork. Joanna Pettet as Detective Janet Baines (1983) Danielle Brisebois as Mary-Frances Sumner #1 (1983–1984) Laurence Haddon as Dr. Mitch Ackerman (1984–1985) The Dr. that delivered Valene's twins and then helped with the ruse that they had died. He later killed himself when confronted by Karen and Mack. Hunt Block as Peter Hollister (1985–1987) Jill Bennett's brother who was killed accidentally by his former fiance Paige. Wendy Fulton (actress)Wendy Fulton as Jean Hackney (1986–1987) Brian Austin Green as Brian Cunningham #2

(1986–1989, 1997) Joshua Devane as Young Greg Sumner (1986–1987; 1990) Lar Park Lincoln as Linda Fairgate (1987, 1989–1991) Eric Fairgate's wife who started a relationship with his brother Michael, and later on with her boss Greg Sumner. She was brutally murdered. Lynne Moody as Pat Williams (1988–1990) wife of Frank Williams and mother of Julie. She and her family were in the Witness Protection Program, but she died when she was hit by Danny Waleska's car. Paul Carafotes as Harold Dyer (1988–1990) Olivia's husband. Peter Reckell as Johnny Rourke (1988–1989) singer and ex lover of Paige Mathison. Kent Masters-King as Julie Williams (1988–1991) teenaged daughter of Frank and Pat Williams. Melinda Culea as Paula Vertosick (1988–1990) Robert Desiderio as Ted Melcher (1988–1989) Betsy Palmer as Aunt Virginia "Ginny" Bullock (1989–1990) Valene's aunt. Sam Behrens as Danny Waleska (1989–1990) Valene's third husband. Joseph Gian as Det. Tom Ryan (1989–1991, 1993) Lorenzo Caccialanza as Nick Schillace/Dimitri Pappas (1990–1991, 1992–1993) Bruce Greenwood as Pierce Lawton (1991–1992) Guest starsKaren Allen (1979) (as Annie Fairgate) Patrick Duffy (1979–82) (as Bobby Ewing) Charlene Tilton (1980) (as Lucy Ewing) Larry Hagman (1980–82) (as J.R. Ewing) Helen Hunt (1980) (as Betsy) Gary Sinise (1980) (as Lee Maddox) Mary Crosby (1980) (as Kristin Shepard) Eve McVeagh (1982) (as Mrs Green) Howard Duff (1984–85, 1990) (as Paul Galveston) Ava Gardner (1985) (as Ruth Galveston) Dick Sargent (1985) (as Himself) Ruth Roman (1986) (as Sylvia Lean) Doug Savant (1986–87) (as Young Mack MacKenzie) Michael York (actor)Michael York (1987–88) (as Charles Scott) Red Buttons (1987) (as Al Baker) Stuart Whitman (1990) (as Willis #2) Ren Hanami (1990) (as Receptionist) Halle Berry (1991) (as Debbie Porter) Marcia Cross (1991–92) (as Victoria Broyard) David James Elliott (1991–92) (as Bill Nolan) Lance Guest (1991) (as Steve Brewer) Darby Hinton (1991) (as police officer) Mary Lou Retton (1992) (as Herself) Billy Bob Thornton (1992) (as a Logger) Plot summariesSeason One (1979–80) 13 Episodes (1–13) Produced by: Joseph B. Wallenstein Knots Landing title card (seasons 1 and 2) Upon arriving in California, Gary and Val (Ted Shackelford and Joan Van Ark) met their neighbors: Sid and Karen Fairgate (played by Don Murray (actor)Don Murray and Michele Lee), the self-described "neighborhood Brady Bunch." Together, they had three teenaged children, Eric (Steve Shaw (actor)Steve Shaw), Diana (Claudia Lonow) and Michael (Patrick Petersen). Sid's daughter from his first marriage, Annie Fairgate who goes and has sex in Karen's and Sid's bed (Karen Allen) would make her only appearance on the Pilot (Knots Landing)pilot episode of the show, and Sid's ex-wife and Annie's mother, Susan Philby (Claudette Nevins) would make her only appearance later in first season episode "Civil Wives"; neither showed for Sid's funeral early during the third season. Likewise, Gary and Val's daughter Lucy (Charlene Tilton) made her only appearance during the first season but then remained on Dallas, even though Gary and Val's original plan was to bring Lucy to California. Also living in the cul-de-sac were corporate lawyer Richard Avery and his wife Laura (John Pleshette and Constance McCashin), parents of a young son, Jason, and young newlyweds Kenny and Ginger Ward (James Houghton and Kim Lankford). In the show's first episode, which aired on December 27, 1979 on CBS, a giddy Valene ran on the beach, proclaiming to Annie Fairgate that she'd never seen the ocean. She would return to the ocean many times over the years, especially after her babies were kidnapped and declared dead during the sixth season. Gary took a job at Knots Landing Motors, a Ford-Lincoln-Mercury (hence, the show's early sponsors) car dealership owned by Sid. He did such a good job that by the end of the first season he was promoted. Gary got drunk at his promotion party, causing a binge drinkingbinge in the two-part season finale, "Bottom of the Bottle". Ginger caught her record producer husband, Kenny, cheating on her with a singer, Sylvie (Louise Vallance), and broke up with him. When Richard wanted to borrow money from Laura's father Hank Murphy (Harry Bellaver), Laura allowed him on the condition that Richard let her go to work. Richard was forced to relent, and Laura saw that she did not have to live by her husband's rules all the time. Season Two (1980–81) 18 Episodes (14–31) Produced by: Joseph B. Wallenstein Knots Landing came back in season two with the introduction of Sid Fairgate's sister Abby Fairgate CunninghamAbby (Donna Mills). Abby was 14 years his junior, and he'd always been oblivious of her scheming ways, but Karen was suspicious of almost everything Abby did. When Abby took a job at Knots Landing Motors, she helped Gary out of a bind with mobsters by going to his brother J.R. Ewing (Larry Hagman). Abby went out of her way to make sure Val found out Gary was cheating on her with Judy Trent (Jane Elliott (actor)Jane Elliott), the attractive wife of Earl Trent, a man he was sponsoring in Alcoholics Anonymous. Even though Abby set her sights on Gary, she settled for Richard, taking no great effort to conceal the affair from Laura, who took on a career in real estate and began a flirtation with her boss, Scooter Warren (Allan Miller). Abby dumped Richard when her ex-husband Jeff Cunningham (Barry Jenner) threatened to take her children away. Viewers discovered exactly how manipulative Abby could be when the women of Knots Landing held a baby shower for Ginger and were interrupted by criminals who held them hostage; Abby seduced one of the thugs, helping to save them.Ginger began dating the father of one of her students while separated from Kenny, but took Kenny back upon discovering that she was pregnant.Sid and Abby's nephew Paul Fairgate (Kale Browne) made his only appearance on the show in the episode "Players". When Sid discovered the car parts Gary and Abby had been buying were actually stolen, his brakes were tampered with in order to keep him from testifying to this in court. Season Two ended with Sid's car going off a cliff, with Sid and his FBI agent bodyguard Salmaggio inside. Season Three (1981–82) 22 Episodes (32–53) Produced by: Joseph B. Wallenstein After Sid's vehicle went off the cliff, he was extricated from his vehicle after it was spotted by a group of teens who witnessed the accident. Salmaggio was pronounced dead at the scene. Sid told paramedics he could not move, and they suspected he had a broken neck. Sid was then taken to Beachview Hospital's intensive care unit. Though initially feared to be paralyzed, Sid showed movement in his limbs, albeit somewhat weak. However, a blood clot rendered him paralyzed from the neck down later in the day. Sid asked Karen to sign papers allowing him to undergo life-risking surgery in order to regain the use of his legs. After some argument, Karen very reluctantly agreed to sign. Sid died in the early stages of the operation, leaving Karen a widow with three children. Karen's unwillingness to discuss her husband's death distanced her from her children, especially from her elder son Eric (Steve Shaw (actor)Steve Shaw), who was angry after she decided to sell Sid's classic hot rod shortly after his death. She took over Knots Landing Motors, firing Abby, but not Gary. Karen would be forced to hire Abby back because Abby's style of bookkeeping could not be easily deciphered. Abby's ex-husband Jeff Cunningham (Barry Jenner) and his mother Annette Cunningham (Carol Bruce) make their final appearance in the episode "The Surprise". Ginger became annoyed as Kenny refused to take interest in her aspirations as a singer. However, both become closer as their first child, Erin Molly, is born, and Karen's dry-eyed composure surrounding Sid's death finally crumbles when Kenny and Ginger put the baby into her arms and ask her to be their new daughter's godmother. Valene attempted to forge a bond with her mother Lilimae (Julie Harris), who moved in with her and Gary. Lilimae had all but abandoned her as a child in her hopes to be a country music singer. Karen's brother, Joe Cooper (Stephen Macht) took on a job as a bookkeeper at University of Southern CaliforniaUSC and briefly stayed with the Fairgate family. Laura began an affair with her boss Scooter after years of being unhappily married to Richard. She planned to leave Richard, but then discovered she was pregnant. She planned on aborting the pregnancy, but decided not to after Richard got violent. Richard was becoming the company pimp after being asked to supply entertainment by his boss for some clients at his law firm, a pattern that kept repeating itself. This was eventually revealed on the news, and Richard's employer denied any cooperation with him. After losing his job, Richard became depressed and had a nervous breakdown. An incident erupted and Richard was sent to a mental institution. After Karen passed up the opportunity to fund a methanol-powered vehicle project, Gary and Abby devoted hours to making it work without her help. Abby made repeated passes at Gary, and finally they slept together. Valene discovered this in the season finale, leaving the cul-de-sac in tears. Season Four (1982–83) 22 Episodes (54–75) Produced by: Peter Dunne With the charges dropped against Sid Fairgate's killers, an angry Karen marched into the office of federal prosecutor M. Patrick "Mack" McKenzie (played by Kevin Dobson). Mack was as upset as Karen, and so they work together to bring the men to justice. They fell in love, and married half-way through the season. Mack's dying father Peter Mackenzie (Jeff Corey) makes his only appearance in the episode "The Block Party". Valene returns to her house in Knots Landing, throwing Gary out. She later becomes a best-selling author with "Capricorn Crude", a thinly veiled expose of the Ewings of Dallas. Gary moves into a beach house with Abby and her two children, and inherits millions from his father's will. Investing money in several ways, he decides to go into the record business and teams up with Kenny who produces the record of a beautiful young singer Ciji Dunne (Lisa Hartman). Upon finding a suicide note at the house, a guilty Laura decided to reconcile with Richard. Richard then proceeded to open a restaurant, Daniel (named after their newborn son). Abby financed the restaurant using Gary's money, and in a cross-promotion, proceeded to demand that Ciji sing there every night. Ginger resented Ciji, as Kenny spent a lot of time worrying about Ciji's career and almost no time at all on Ginger's. Laura and Ciji became best friends, which prompted the insecure Richard to fear the women were having a lesbian relationship. Valene's publicist Chip Roberts (Michael Sabatino) convinced Lilimae to let him live with them. Chip dated Karen's daughter Diana and Ciji at the same time, and panicked when he found out Ciji was carrying his child and after Ciji discovered his true identity, Tony Fenice. In season four, Knots Landing, which had been a top 30 show during its first two seasons and a top 45 for its third, became a huge ratings success. The climax came when Ciji's dead body washed up on the beach. Gary's life collapsed as he once again turned to the bottle to escape his problems. He woke up on the beach on which Ciji's body had been discovered, and his drinking caused him not to remember anything of the night before. He was arrested for her murder. After Ciji's death, Laura refused to forgive Richard for being mean to her best friend while she was alive. Believing Laura no longer loved him, Richard left Knots Landing. Laura thought Richard left town in this manner because he killed Ciji, but later came to know otherwise. Kenny and Ginger also left Knots Landing, with Kenny being offered a job in Nashville, TennesseeNashville. In the season finale, Gary languished in a prison cell awaiting trial for Ciji's murder, while Lilimae discovered Chip's real identity. By this time, Chip had left town with Diana, leaving Karen horrified that her daughter was in the hands of a killer. Season Five (1983–84) 25 Episodes (76–100) Produced by: Peter Dunne In the season premiere, Gary was acquitted of Ciji's murder due to a lack of evidence. Chip (now revealed to be murder suspect Tony Finece) and Diana fled town as Chip became the primary suspect. Meanwhile, Valene began dating reporter Ben Gibson (Douglas Sheehan). Diana married Chip while on the run with him. Even though Chip confessed to Diana that he'd killed Ciji, the state had no case without her testimony. When it looked as though Chip would get away with the murder, Lilimae took matters into her own hands by running him down with her car. However, Chip was merely bruised by the impact. Meanwhile, Abby and Sid's uncle died, leaving Abby and (by default) Karen land inheritance at Lotus Point, California. Abby then formed Apolune Corp., a secret subsidiary of Gary's company, with which to build her power base. Abby convinced Gary to marry her in order to get half of his inheritance. Before marrying Abby, Gary slept with Valene one more time. By the time Valene discovered she was pregnant, she was in love with Ben, but he could not take the fact that she was pregnant with another man's babies and they broke up. Abby hired a Ciji look-alike, Cathy Geary (Lisa Hartman), to keep Gary distracted so he would not learn about her crooked business practices. Gary hired Cathy to work at his ranch, but when Chip (who had been hiding out at the ranch with Diana) saw Cathy, he fell backward onto a pitchfork and killed himself. The ordeal with Diana caused Karen to become more depressed, and she was addicted to prescription painkillers. United States SenateState Senator Greg Sumner (William Devane), an old college friend of Mack's who was running for US Senator, received the endorsements of Mack and Karen, as well as Abby, who wanted to buy herself a senator. Greg hired Mack to lead his Senate Crime Commission investigating the crooked Wolfbridge Group, and when Abby convinced Greg to get her an illegal land Variance (land use)variance to build on Lotus Point, Greg pinned it on Mack, thus ending their friendship. After Karen's recovery from drug addiction, she and Diana reconciled but Diana moved to New York to live near her uncle, Joe Cooper, and to pursue a career. Wolfbridge forced Abby to let them become her partners in the development of Lotus Point. When Mack discovered that Apolune owned all the land in Lotus Point, except Karen and Abby's inheritances, he thought Apolune was a division of Wolfbridge. Laura confessed that Apolune was Abby's company and Gary decided to divorce Abby. When Mack faked Gary's death to get Wolfbridge, Karen left him, believing his obsession with Wolfbridge had cost her everything. In the season finale, Mack tried to trick Mark St. Claire (Joseph Chapman (actor)Joseph Chapman), the head of the Wolfbridge Group, into trying to shoot Gary, but St. Claire's assassin hit Karen instead. Abby was kidnapped by the Wolfbridge Group when they realized they had failed. Season Six (1984–85) 30 episodes (101–130) Produced by: Peter Dunne (ep. 101-119)/Lawrence Kasha (ep. 120-130) Greg Sumner killed Mark St. Claire, who was holding Abby hostage on a boat. After being shot, Karen declined to be operated upon, and was given six months to live. Lilimae's illegitimate son Joshua (Alec Baldwin), whom she'd abandoned as a baby, came to live with her and Valene. He began dating Cathy, who'd moved in with Laura. Abby began running the television station that Ben worked at after Gary bought it. She was shocked to learn that Gary was the father of Valene's babies. Abby told Scott Easton (Jack Bannon), a man working for Lotus Point, about her problem, and he had Valene's babies kidnapped after their birth, though without Abby's consent. Valene was told that the babies were stillborn, and when she insisted that she'd heard them crying, nobody believed her. Heartbroken over the loss, Valene left Knots Landing for Tennessee, where she developed psychosomatic amnesia, calling herself Verna Ellers (a character from one of her novels) and becoming a waitress. Meanwhile, Gary and Abby reconciled and the couple, along with Karen, agreed to become partners in Lotus Point. Karen took Mack back on Gary's insistence after she confided to him that she was dying. Mack convinced Karen to have the surgery correcting her back problem. Paul Galveston befriended Gary, and offered him a chance to develop Empire Valley, a large piece of real estate. Gary began construction. Galveston died, leaving his company to Greg Sumner, Galveston's son. Assuming he would inherit Empire Valley, Greg resigned from the Senate, but Galveston left it to Gary. Greg's mother Ruth Galveston paid him an extended visit. She took a liking to Abby and paid Abby to pretend to have slept with Greg in order to break up his relationship with Laura. Joshua became a televangelist at Abby and Gary's TV station. He and Cathy married, though on their wedding day she had second thoughts because fame had warped him. Valene (still believing herself to be "Verna") became engaged to a man in Tennessee, but Gary finally found her and brought her back to Knots Landing where she re-united with Ben but still refused to believe that her babies were dead. Karen and Ben discovered that Valene's doctor had paid off a nurse to assist in the deceit. Mack tracked down Valene's babies that had been adopted on the black market. In the season finale, Abby and Val, Mack and Ben, and Gary and Karen arrived at the Fisher house, where Sheila Fisher opened the door holding one of Valene's twins in her arms. As Valene got out of Abby's car and approached the house, Harry Fisher passed in front of the house in his car with Valene's other twin baby. His wife shouted to him and warned him, "They want to take the babies..." Harry drove off with one of Valene's babies before anyone could stop him. This season finale, entitled "The Long and Winding Road" , ranked as the #1 program for the week in the Nielsen Ratings. It is the only episode to rank #1 for the week during the series' 14 seasons. Season Seven (1985–86) 30 episodes (131–160) Supervising Producer: David Paulsen Produced by: Lawrence Kasha In the season premiere, Valene waited outside the house of the couple who had her babies. The babies were returned to her in episode 2. Valene and Ben agreed to raise the babies together, and they married. Also married this year were Laura and Greg, whose estrangement ended when he convinced her that his mother was out of their lives for good. Gary received news about his brother Bobby's death, and headed back to Dallas for the funeral (in the 1986-87 season premiere of Dallas, Bobby Ewing's death was revealed to be part of a dream. However, this was never acknowledged on Knots Landing, and in the chronology of this series he remained dead.)Empire Valley went awry when Gary discovered that Galveston Industries and its partners were secretly building an underground espionage operation there. Gary blew up Empire Valley to stop this. Gary blamed Abby, who had helped Greg and gone behind his back. Gary informed Abby at the end of the season that he would divorce her. Gary bequeathed part of Empire Valley to Val's kids Betsy and Bobby (named after his recently deceased brother), who were being raised as Ben's children. Peter Hollister (Hunt Block) went to work for Greg, and Jill Bennett (Teri Austin) became Mack's new colleague. It was later revealed that Peter and Jill were brother and sister, whose family had been swindled by Paul Galveston and wanted revenge. Jill abandoned her side of the plan when she fell in love with the now-single Gary, while Peter, in going through Greg's files, discovered a woman named Sylvia Lean (Ruth Roman) who'd been receiving monthly payments from Galveston Industries for over 20 years. She had given birth to Galveston's now-deceased son, whom Peter pretended to be, demanding compensation from Greg. Abby discovered the truth and demanded 51% of Peter's compensation in return for keeping it secret, but when Peter discovered that Abby had known that Valene's babies had been kidnapped from the start, he used this against her. With his sanity slipping away, Joshua became abusive toward Cathy, who then left him. Not wanting to lose her, he decided that they should die together. He tried to throw Cathy off of a tall building, but Lilimae arrived in time to stop him. She disowned Joshua, but he fell from the roof and died. Lilimae spent the rest of the season mourning him, while Cathy moved on to Ben, whose marriage to Valene was failing due to the continued threat of Gary, who by then knew that he was the father of her twins. Ben cheated with Cathy and in the season finale, prepared to leave Valene and go on a singing tour with her as her manager. Abby's teenaged daughter Olivia (Tonya Crowe) developed a drug problem. A girl, Paige Matheson, (Nicollette Sheridan) arrived at Mack's door claiming to be his daughter. But on the same day, Karen disappears and finds herself being held hostage by a stranger who knows her name. Season Eight (1986–87) 30 episodes (161–190) Produced by: Lawrence Kasha Karen's kidnapper was Phil Harbert (Louis Giambalvo), an old friend of Greg's and Mack's who was bitter toward Mack, whom he blamed for having caused the death of his wife. Mack suspected Greg for the kidnapping, but believed that Greg had nothing to do with it. Ben returned to Valene and promised to make their marriage work, however this promise was compromised when he was contacted by Jean Hackney (Wendy Fulton), who had tricked Ben into doing a mission for an underground organization to kill Greg Sumner, and when Ben and Valene's children were threatened, he finally agreed to do this. Luckily, he did not succeed, although Jean Hackney's insistence on following him meant that he had to flee the country and leave Valene at the end of the season. Laura refused to forgive Valene for Ben's plan to kill Greg and ended their friendship. Finally, Peter is Jill's sister and Greg's brother. Laura announced to Greg that she was pregnant. Mack had a difficult time handling Paige, who moved in with the Mackenzies and seduced Karen's son Michael. When Karen discovered Paige Matheson was dead, Paige told the Mackenzies she'd faked her own death in order to escape her grandparents. Paige's mother (and Mack's first love) Anne (Michelle Phillips) arrived in Knots Landing to be with her daughter, whom she'd believed dead. Anne moved in to Laura's house on the cul-de-sac. Anne attempted to seduce Mack, but failed. She left Knots Landing. Abby's daughter Olivia turned to drugs. When her drug dealer beat up her younger brother Brian, she gave all her drugs to Abby, who encouraged her daughter's friendship with Peter Hollister. Olivia's friendship turned to crush. Peter was only interested in Paige. At the end of the season, Abby found Olivia standing over Peter's body. They both assumed the other had killed him. She hid Peter's corpse under the soil where cement was to be poured the next day. In the final scene of the season, Karen informed Abby that there was a crack in the cement, and it would have to be re-poured. Season Nine (1987–88)29 episodes (191–219) Produced by: Lawrence Kasha/Mary-Catherine Harold A melancholy Laura extended the olive branch to Valene at Karen's insistence. Paige had accidentally caused Peter's death. No charges were brought against her, but Abby swore that she would make Paige pay for framing Olivia for the crime. Laura told Greg that she was dying of cancer. Greg refused to accept it, and Laura left him to go to a clinic, where she died. Laura's funeral brought the return of Richard, who moved his and Laura's sons back to the east coast with him and his new wife. Greg was left to raise Meg, his daughter with Laura. Feeling unable to raise the child by himself, Greg offered to allow Karen and Mack to adopt her. Meanwhile, Abby rekindled an old romance with wealthy business executive Charles Scott (Michael York (actor)Michael York) and the two married briefly. But Abby ended the relationship after discovering his ulterior motives about plans for the Lotus Point Marina. After Laura's death, Frank and Pat Williams (Larry Riley (actor)Larry Riley and Lynne Moody) and their daughter Julie (Kent Masters-King) moved onto the cul-de-sac. Mack and Karen were suspicious of them at first, but became their friends upon discovering they were in the Witness Protection Program. Business at Lotus Point began to slow down, and Karen, Gary, and Abby agreed to expand the marina to accommodate the business of Manny Vasquez (John Aprea). Meanwhile, Paige, Michael, and Paige's boyfriend Johnny Rourke (Peter Reckell) went on an archaeological dig in Mexico. When it became apparent that Manny Vasquez (John Aprea) was selling drugs through Lotus Point, its owners tried to stop him. He told them that Paige and Michael would not re-enter the United States until he was allowed to conduct business at Lotus Point. Paige began to work for Greg, and found herself attracted to him. Lilimae left Knots Landing with a new man in her life. Gary took an increased interest in Valene's twins, initially at Valene's resistance. His attention to Val and the children was an annoyance to Jill, who was now living with Gary and became jealous. Jill made an elaborate plan where she edited soundaudio recordings of Ben's voice to make Valene think Ben was trying to contact her, and forged letters from him. Jill then stole the tapes of Ben's messages and the letters, leaving Valene with no proof when she tried telling people. In the season finale, Jill broke into Valene's house at night and forced Valene at gunpoint to take a whole bottle of sleeping pills in order to make it look like Valene had killed herself. The following morning, Valene was seen laying motionless on the floor. Season Ten (1988–89) 28 episodes (220–247) Co-Executive Producer: Lawrence Kasha Produced by: Mary-Catherine Harold/Lynn Marie Latham/Bernard Lechowick Frank found Valene unconscious and phoned for an ambulance. Valene survived and said that Jill had tried to kill her, but Jill had provided herself with an alibi; she slipped drugs into a man's drink and brought him to her hotel room in San Francisco. He passed out, but as she was there when he awoke in the morning, he assumed she'd been there all night. When the man got an Sexually transmitted diseaseSTD and Jill and Gary's tests turned up negative, Gary investigated and found that the man remembered nothing of the night he had supposedly slept with Jill. With doubt on her innocence, Jill was left friendless and she tied herself up and put herself in Gary's trunk so he'd be framed for trying to kill her. However, Jill actually died and Gary was suspected of killing her. Mack was able to get the charges dropped (as he had done with Paige a season earlier) when a hidden fingerprint proved Jill closed the trunk of Gary's car herself and the whole thing was exposed as a plot by Jill to frame Gary.Valene's Aunt Virginia "Ginny" Bullock (Lilimae's sister) came to Visit Knots Landing and ended up moving in with Valene on Sea-view Circle. A con artist tried to extort money from Frank and Pat by threatening to reveal their location to the men who had threatened Pat's life when she testified against them. Mack halted the con artist, and used dental records from a deceased family to make the men believe that the Sollars family (the original identity of the Williamses) were dead. The young crowd was able to get out of Mexico, and Manny Vasquez's nephew Harold Dyer (Paul Carafotes) (who was also dating Abby's daughter Olivia) killed his uncle when Manny tried to kill Mack. Paige seduced Greg and the two began dating. Unfortunately, Greg wanted to become mayor of Los Angeles and he realized the much-younger Paige would not make a good political wife. He asked Abby to marry him instead, and Abby got her revenge against the broken-hearted Paige, who still kept her job at Greg's company nonetheless. Mack had another mid-life crisis and quit his job, opening a private practice. Following a health and safety report, the Lotus Point resort has to be sold. However, Abby had discovered that there was oil underneath Lotus Point, and created a company to conceal who was buying the resort from her partners. Paige discovered that the whole thing was a scam initiated by Abby and her cohort Ted Melcher which infuriated Karen who was now intent on sending Abby to jail for fraud. Abby narrowly avoided the fraud charges and agreed to donate Lotus Point a public park. She then left Greg (and Knots Landing) for a Trade Envoy job in Japan (a job Greg himself had planned to get after his failed attempt to become mayor of Los Angeles). Season Eleven (1989–90)29 episodes (248–276) Co-Executive Producer: Lawrence Kasha Produced by: Mary-Catherine Harold/Lynn Marie Latham/Bernard Lechowick Valene and Gary became separately involved with another divorced couple, Danny and Amanda Waleska. Danny (Sam Behrens) had beaten and raped Amanda (Penny Peyser). Valene proclaimed that she loved Danny more than she'd ever loved Gary or Ben. As Danny became more and more abusive and distant, Valene realized the truth, and kicked him out of her house. Paige fell in love with Tom Ryan (Joseph Gian), a dirty cop. The two planned on getting married, but on the day of their wedding, Tom ran away after Greg blackmailed him. Danny hit Pat with his car and then drowned in the Williams' swimming pool after trying to attack Julie. Gary proposed to Valene, who accepted. Karen started presenting a television talk show, but was stalked by one of her producers, who had been put up to it by another producer Dianne Kirkwood (Robin Strasser) who hated Karen. Olivia married Harold, but they had monetary woes due to Abby's decision to cut Olivia off financially. They left the show mid-season. Karen's son Eric left his wife Linda at the Mackenzie household. Linda had an affair with Eric's brother Michael, then divorced Eric. Paige's mother Anne returned to Knots Landing penniless, and took up with Greg. Greg's daughter Mary Frances (Stacy Galina) visited him after a 6-year estrangement and was later shot dead in Greg's office tower. The bullet was meant for Greg. Season Twelve (1990–91) 27 episodes (277–303) Co-Executive Producer: Lawrence Kasha Produced by: Mary-Catherine Harold/Lynn Marie Latham/Bernard Lechowick Frank had Pat's life support disconnected. New to Knots Landing were Greg's sister Claudia (Kathleen Noone) and his niece Kate, who shared an uncanny resemblance to dead Mary Frances (Kate was also played by Stacy Galina). Kate met Steve Brewer (Lance Guest), a man claiming to be her brother, at school. Steve, who had been put up for adoption by Claudia, was the product of a tryst between Claudia and Paul Galveston. Claudia planted a gun on former felon Steve and he fled from the police. Mack befriended a high schooler named Jason Lochner (Thomas Wilson Brown), who was being beaten by his father. Mack recalled the memories of his own physical abuse at the hands of his now late father, and invited Jason to live with him and Karen. Jason dated Julie Williams for a while. Valene's Aunt Ginny Bullock (Betsy Palmer) left Knots Landing for Europe. Valene and Gary got together after Danny's death. Valene fell from a horse and developed psychiatric problems for a while. After a season of irrational behavior, Valene was cured, and she married Gary again. Anne met Italian Nick Schillace (Lorenzo Caccialanza), with whom she both fell in love and spent the season scheming. Their plans all backfired, and in the season finale, Anne was left broke and homeless. Paige and Linda competed at the Sumner Group (the renamed Galveston Industries). Season Thirteen (1991–92) 22 episodes (304–325) Co-Executive Producer: John Romano (ep. 304–318) Supervising Producer: Joseph Hardy Senior Producer: Ann Marcus (ep. 319–325) Produced by: Mary-Catherine Harold Staff Writers ("story editors"): novelist Rachel Cline ("What to Keep," "My Liar"), novelist James Magnuson ("The Money Tree"),playwright and television producer Donald Marcus, and television writer Lisa Seidman. Kate blamed Claudia for Steve's death and cut her mother out of her life, but they reconciled after Claudia took an overdose of pills. After Nick left her, Anne was left homeless and bankrupt. She started an advice show on night-time radio which became a smash hit. Linda was murdered by the crazy Brian Johnston, who held the Mackenzies hostage for an entire episode. Jason left the Mackenzie household for Sweden while Julie left Knots Landing. Frank dated Debbie Porter (played by Halle Berry). Gary invested in a project that turned energy from the ocean's tides into usable electric energy. The man behind it, Joseph Barringer (Mark Soper), became Kate's boyfriend. Paige joined Gary, along with her new boyfriend Pierce (Bruce Greenwood). A former flame of Pierce's, Victoria Broyelard (played by Marcia Cross), revealed that Pierce had killed his previous girlfriend. Paige did not believe it. Paige was shot by Pierce (aiming for Greg, whom he hated) and temporarily paralyzed, giving a delusional Pierce (who thought she was the previous, pregnant girlfriend he'd already killed) a chance to kidnap her. Mack and Greg rescued Paige and Pierce was arrested. Alex Barth (Boyd Kestner), the nephew of the Galvestons' housekeeper, blackmailed Claudia for refusing to help her ailing mother with her medication and thus allowed her to die. Claudia said that she wanted to end their mother's pain. Claudia disliked Joseph and asked Greg to get him a job away from Kate. Gary, who had tied up all his money in Tidal Energy, was left bankrupt. Valene attempted to console him, and the two of them spent the rest of the season enjoying a happy marriage. Valene was to write a biography on Greg Sumner. Mary Robeson (Maree Cheatham) told Valene that she was Meg's grandmother, as she was Laura's biological mother. Valene refused to believe it, and when she investigated Mary Robeson further, she was kidnapped in Florida. Season Fourteen (1992–93) 19 episodes (326–344) Co-Executive Producer: Barbara Corday Supervising Producer: Ann Marcus Produced by: Mary-Catherine Harold Gary believed Valene dead when he saw a car with her in it blow up. Greg attempted to retire from the Sumner Group, leaving one-third to Claudia, one-third to Paige, and one-third to Mack and Karen (in a trust for Meg). Anne's pregnancy prompted him to return to Knots Landing to marry her. Anne discovered she had suffered from hysterical pregnancy. Anne longed for Nick, who had returned to Knots Landing. When Mack sought $1,000,000 to bribe Mary Robeson to stop seeking custody of Meg, Claudia gave him $500,000 and gave the rest to Nick to start a restaurant. Paige demanded an audit, and Claudia had to ask Nick for the $500,000 back. Nick had already spent it and got the money elsewhere. It turned out a man named Nigel Treadwell (Daniel Gerroll) and his mysterious partner had been trying to take over the Sumner Group with this $500,000. Karen did not believe Mack would do something like that and left him, staying with Diana in New York City. In the two-hour series finale on May 13, 1993, Karen returned to Mack, and Nick's accomplice Vanessa Hunt (Felicity Waterman) killed Treadwell, paving the way for Valene to return to Gary. Treadwell's partner was revealed to be Abby, who, returning for the first time in four years, told Greg that she would be taking over the Sumner Group. However Greg thwarted her scheme by threatening to reveal her dirty dealings in Japan with some well documented evidence that he had just acquired. Anne offered Greg a divorce and he was free to get back together with Paige. Claudia decided to move to Monaco, and on the way to the airplane met up with Anne and Nick. Back at Seaview Circle, Gary, Val, Mack and Karen were preparing to have a barbecue as a new couple were moving into Frank's old house on the cul-de-sac. At that moment, another car pulled up and Abby emerged. She informed everyone that she had heard Claudia's house was for sale (the house which had originally been hers when she lived on the cul-de-sac). Valene smiled and welcomed Abby back to the cul-de-sac before rushing Gary away. Abby is left with Karen, to whom she proclaimed "Just like old times, isn't it?" Knots Landing/Dallas crossover episodesBetween Seasons one and four, there were nine episodes where Dallas characters appeared, played by their respective actors. Season 1

Episode 1: "Pilot". Guest starring Patrick Duffy as Bobby Ewing. Episode 2: "Community Spirit". Guest starring Larry Hagman as J.R. Ewing. Episode 6: "Home is For Healing" Guest starring Charlene Tilton as Lucy Ewing. Season 2 Episode 5: "Kristen". Guest starring Mary Crosby as Kristen Shepard. Episode 9: "A Family Matter". Guest starring Larry Hagman as J.R. Ewing. Episode 13: "The Loudest Word". Guest starring Patrick Duffy as Bobby Ewing. Episode 17: "Designs". Guest starring Larry Hagman as J.R. Ewing. Season 4 Episode 2: "Daniel". Guest starring Larry Hagman as J.R. Ewing. Episode 6: "New Beginnings". Guest starring Larry Hagman as J.R. Ewing, Patrick Duffy as Bobby Ewing and Eric Farlow as Christopher Ewing. Dallas (2012 TV series) With the Dallas (2012 TV series)2012 continuation of Dallas, there was talk of some Knots Landing characters returning to air. Rumors began surfacing that both Van Ark and Shackelford would reprise their roles as Valene and Gary. Ted Shackelford passed on the small role the producers offered him to reprise his role as Gary Ewing on the new series. Van Ark refuted the claims, saying: "Well, rumor has it… I'm having lunch with Ted tomorrow. They asked him to come down and be part of it. It was several months ago, and he's deeply into filming The Young And The Restless, the daytime soap for CBS. But if they asked Gary, maybe they'll ask [for Val]."Keck, William (June 4, 2012). "On the Set: The Ewings Return to Dallas". Seattle Post-Intelligencer. Hearst Corporation. . Retrieved June 24, 2012.Byrne, Craig. "TV Out Of The Box: Joan Van Ark On Whether We'll Be Seeing Valene On Dallas". KSiteTV.com. . Retrieved June 6, 2012. . "There is no doubt about it. This man I loved so much, Larry Hagman… I was just at his birthday party several months ago… I would kill, that if Val came back, if it was trouble for J.R. Because they were always at odds." However, Gary and Valene's daughter, Lucy Ewing, along with Ray Krebbs have made appearances in the new series. Although both Shackleford and Van Ark appeared in the 1991 series finale of Dallas playing "alternative" versions of Gary and Valene (the episode depicted a fantasy world in which J.R. had never been born), there have been no real crossovers of story or characters from Knots Landing to Dallas or vice-versa since the 1985-86 season of each show (which, on Dallas, turned out to be a dream had by Pam Ewing).It was announced in 2012 that Ted Shackelford and Joan Van Ark will return to [Dallas] in 2013 for the second season. Behind the scenes The actors had more input than actors on other 1980s primetime soaps. In 1987, the writers wanted Mack (Kevin Dobson) to have an extramarital affair with Anne (Michelle Phillips). Michele Lee, who played Mack's wife Karen, protested this to David Jacobs, saying, "There has to be one stable couple on the show." The extramarital affair storyline was nixed, and Michelle Phillips, who had been signed to a contract, was written out for a few seasons before returning in 1990. When she did return, Anne did not pursue Mack. William Devane, who played Greg Sumner, re-wrote most of his character's dialogue, to the point where, in co-star Michele Lee's words, "most people (on set) were (probably) frightened of him." The Gary/Val/Abby triangle that provided story throughout the mid-1980s was suggested by Ted Shackelford and Joan Van Ark in 1980, and the producers hesitated for a year and a half before going through with it in 1982. The famous 1984 storyline where Valene's babies got kidnapped was originally envisioned as one of scheming Abby's plots. Donna Mills, who played Abby, acknowledged that her character was evil but did not think she was that evil. Fearing the audience would never forgive her character for kidnapping another woman's babies, she asked the writers to make the kidnappings a result of Abby's actions, but only by accident, and the writers complied. The writing team of Bernard Lechowick and Lynn Marie Latham (the head writers from 1986 to 1991) was controversial among both fans and actors. Their humor-imbued style of writing made them the favorites of Michele Lee, while John Pleshette felt they were "awful people." Pleshette, however, was not a cast member during their tenure and harbored resentment because the writing team, who had been represented by his wife, moved to a different agency. Joan van Ark, whose character was struck by a brain illness in season 12 and proceeded to thereafter go crazy, felt that Latham and Lechowick had turned her character into the "village idiot." Joan Van Ark and Donna Mills' favorite Knots Landing writer was Peter Dunne, who was responsible for making Knots Landing a top ten show in 1984.In 1987, CBS demanded that production costs be cut. This meant the firing of two regulars, Constance McCashin and Julie Harris. Season 13 saw a large ratings drop for the show after writer/producers Bernard Lechowick and Lynn Marie Latham left to create Homefront (US TV series)Homefront and creator David Jacobs had a health crisis and pulled back his involvement in production. Jacobs has publicly stated that the way he knew the show was in trouble was when waitresses at his favorite diner, whom he had heard gossiping about Knots Landing every Friday during past seasons, suddenly stopped discussing the show in late 1991. He attempted to save face by shutting down production on November 20, 1991 , firing head writer John Romano, and replacing him with Ann Marcus. Cost cutting again plagued the series in its final season, when only 19 episodes were produced, and regular characters did not appear in each episode. In one episode in the final season ("My Kingdom for a Horse"), only three of the fourteen regulars cast for season 8 appeared (Devane, Phillips, and Lee.) Not wanting to compromise what he felt had been a good run, series creator David Jacobs described its end as a "mutual decision" between Knots Landing's producers and the CBS Network, saying, "We don't know if they would have picked us up anyway...but even if they had, we would have had to pare away more to survive."Music The theme song, which lasted all 14 seasons was composed by Jerrold Immel. Immel, along with Craig Huxley composed the background music for the Pilot (Knots Landing)pilot. The original background music cues by Immel and Huxley were never fully abandoned by the show, and were heard as late as the final season. The early Knots Landing background music cues heavily emphasized the brass section, and were often played with a very sparse bass line accompaniment. It was, in fact, the only aspect of the series ever to win an Emmy award, for the music orchestration during its 1979–80 season. By season 4 of Knots Landing, the lushness of the 1980s was in full swing and Knots Landing's background cues reflected that style. The new dramatic cues emphasized full orchestral arrangements as formerly middle class Knots Landing became upwardly mobile. The background music of seasons 4–7 was frequently composed by either Lance Rubin or Ron Grant (composer)Ron Grant. Season 8 introduced a completely new score for Knots Landing. By 1986 New Wave musicNew Wave artists and bands had taken America by storm and the new style of music cues made good use of the synthesizer instead of a full orchestra. Bruce Miller (composer)Bruce Miller was one of the main composers during this era. Updated orchestrations of the by-then-familiar Lance Rubin cues were also re-arranged to be played by the synthesizer, and the Immel/Huxley cues were similarly utilized, albeit less commonly.In the early 1990s, soft contemporary acoustic music became popular and Knots Landing began incorporating this into its background music during season 12. Lance Rubin's music cues were completely phased out at this point. Patrick Gleeson and Kennard Ramsey composed during this period.Opening creditsKnots Landing had five completely different styles of opening credits over its 14 years, in contrast to Dallas (1978 TV series)Dallas, which changed only to a variation on its original style for its final two seasons. (December 27, 1979 – March 26, 1981) The original opening of Knots Landing designed by Wayne Fitzgerald features a spinning overhead shot of a California beach which dissolves to a spinning skyview of roads and houses, gradually zooming in to a freeze-frame of the Seaview Circle cul-de-sac. The camera then zooms in to the top of each house in turn, showing a brief shot of the couples who reside in each house. Photo credits for each of the main actors then appear in a shape superimposed over the cul-de-sac. This title sequence was used only during seasons 1 and 2.(November 12, 1981 – May 14, 1987) Knots Landing unveiled a new opening at the start of season 3. In what is probably the best-remembered introduction, the sequence designed by Gene Kraft features the title scrolling from right to left followed by a parade of clips of the show playing in boxes. Each cast member is credited below a big box showcasing a close-up of their character, accompanied by three or four smaller boxes showing that character in a scene with other characters. (September 24, 1987 – May 18, 1989) At the beginning of the ninth season, Knots Landing's producers decided to break tradition with the opening. The intro designed by Sandy Dvore now features a slow panning over a painting similar to the splattered style of Jackson Pollock. As the zig-zag panning continues, the cast montage appears, featuring black and white shots of the actors inside of small ovular cameos. The posed cameos were dropped in the 1988–89 season in favour of color close-ups taken from the show. (September 28, 1989 – May 16, 1990) The eleventh season of Knots Landing saw the show unveil its fourth title sequence. The new opening designed by Castle/Bryant/Johnsen now showcases sandcastle structures of the cul-de-sac houses as well as some skyscrapers representing Los Angeles on a beautiful beach. It is the only version of the opening without pictures of the cast. The camera twists through this sandcastle community with only the actors' names appearing. Another and never-seen particularity appears right from the second episode of this season: the opening splits in two. The cutting takes place during the staff credits. The main theme is adapted as well. Between the two parts of the opening is presented a summary of what happened in the past episodes. (September 13,

1990 – May 13, 1993) To redefine the show for the 1990s, Knots Landing made one final change to the opening credits by returning to its famous across-the-screen style of scrolling clips. This version was designed by Castle/Bryant/Johnsen to be "safe" for the 1990s. Eventually it would bring Knots Landing through its final three seasons on the air. Nielsen Ratings Season Premiere Finale Episodes Timeslot Rank Households(in millions) Season 1 December 27, 1979 March 27, 1980 13 Thursday 10:00 pm #30 20.0 Season 2 November 20, 1980 March 26, 1981 18 #28 19.0 Season 3 November 12, 1981 May 6, 1982 22 #43 15.3 Season 4 September 30, 1982 March 10, 1983 22 #20 18.6 Season 5 September 9, 1983 March 29, 1984 25 #11 20.8 Season 6 September 20, 1984 May 16, 1985 30 #9 20.0 Season 7 September 19, 1985 May 18, 1986 30 #17 16.7 Season 8 September 18, 1986 May 7, 1987 30 #26 16.8 Season 9 September 24, 1987 May 12, 1988 29 #31 15.9 Season 10 October 27, 1988 May 18, 1989 28 #25 16.2 Season 11 September 23, 1989 May 11, 1990 29 #34 14.3 Season 12 September 13, 1990 May 16, 1991 27 #35 13.74 Season 13 September 12, 1991 April 9, 1992 22 #41 12.32 Season 14 October 29, 1992 May 13, 1993 19 #39 11.68 RerunsKnots Landing was first packaged into syndication for off-network broadcast in local markets by Lorimar in 1984. This package started with the "Lorimar Presents" logo and had a narrated recap of each episode set to the theme music. The first 190 episode were packaged this way, but only the first 160 episodes were seen in local market syndication. Knots Landing then appeared on the cable channel TNT in September 1992 and ran through May 1993 (only the first 160 episodes were broadcasted during this run.) TNT became the first network to broadcast all 344 episodes of Knots Landing during its second syndication run to the series beginning in 1995. SoapNet later acquired the rights to the series when it first went on the air in January of 2000. Knots Landing has not been seen in United States syndication since SoapNet quit broadcasting it.DVD releasesThe first season of Knots Landing was released on DVD on March 28, 2006 in Region 1. Fans of the series lobbied Warner Home Video via an online petition at www.knotslanding.net for further releases, and Warner Home Video released Season Two on April 14, 2009. There are currently no plans for any further seasons being released due to unsatisfactory sales of the first two seasons.DVD Season Ep # Region 1 Region 2 Region 4 Comments Season 1 13 March 28, 2006 February 19, 2007 June 6, 2007 The first-season DVD box set has five single-sided discs. The Region 1 and 2 releases include commentaries by actors Ted Shackelford and Joan Van Ark. The set also includes a featurette/clip from the 2005 Together Again non-fiction reunion show in which stars Ted Shackelford and Joan Van Ark reminisce about the series. Season 2 18 April 14, 2009 October 16, 2009 (Germany) TBA The second-season DVD box set has four single-sided discs. The season introduces Donna Mills as Abby Cunningham, and Larry Hagman returns to guest star as J.R. Ewing during this season. There is no bonus material.Who lived where The series' signature cul-de-sac, Seaview Circle, was actually Crystalaire Place in Granada Hills, California, a suburban street in Los Angeles' San Fernando Valley about 20 miles north of the Pacific Ocean. The opening credits during the first two seasons were edited in such a way to make it appear that the cul-de-sac was closer to the beach. 16961 Seaview CircleKenny Ward (Seasons 1–4) Ginger Ward (Seasons 1–4) Erin Molly Ward (Seasons 3–4) 16969 Seaview CircleAbby Cunningham (Seasons 2–4, 14–reunion) Brian Cunningham (Seasons 2–4) Olivia Cunningham (Seasons 2–4) Gary Ewing (Season 4) Claudia Whittaker (Seasons 12–14) Kate Whittaker (Season 12) Alex Barth (Season 13) 16975 Seaview CircleRichard Avery (Seasons 1–4) Laura Avery (Seasons 1–8) Jason Avery (Seasons 1–8) Daniel Avery (Seasons 4–8) Cathy Geary (Seasons 6, 7) Anne Matheson (Season 8) Patricia Williams (Seasons 9–11) Frank Williams (Seasons 9–13) Julie Williams (Seasons 9–13) 16972 Seaview CircleKaren Cooper Fairgate Mackenzie (Seasons 1–14, reunion) Sid Fairgate (Seasons 1–3) Diana Fairgate (Seasons 1–4) Eric Fairgate (Seasons 1–9) Michael Fairgate (Seasons 1–12) Joe Cooper (Seasons 3-4) Mack Mackenzie (Seasons 4–14, reunion) Mary Frances Sumner (Season 5) Paige Matheson (Seasons 8–10, 13) Linda Fairgate (Season 11) Meg Mackenzie (Seasons 9–14, reunion) Jason Lochner (Seasons 12–13) Abby Cunningham (reunion) 16966 Seaview CircleGary Ewing (Seasons 1–3, 13–14, reunion) Valene Ewing (Seasons 1–13, 14–reunion) Lilimae Clements (Seasons 3–9) Chip Roberts (Season 4) Joshua Rush (Seasons 6–7) Cathy Geary (Seasons 6–7) Bobby Ewing (Seasons 7–14, reunion) Betsy Ewing (Seasons 7–14, reunion) Ben Gibson (Seasons 7–8) Aunt Ginny (Seasons 10–12) Danny Waleska (Seasons 11–12) See alsoList of Knots Landing episodesThe game World of Warcraft, which features a coastal resort town called "Schnottz's Landing" which spoofs the series. 1FootnotesSources Roush, Matt. "Series to End After 14 Years". USA

Today. January 11, 1993.Nielsen Rating Information from the following source: Brooks, Tim and Marsh, Earle. The Complete Directory to Prime Time Network and Cable TV Shows 1946–Present: Eighth Edition. 2003. Random House. New York. Exclusive web Interviews ©KnotsLanding.net 2006 Some actors quotations from the following website: http://www.knotslandingonline.com/. Interviews ©Arthur Swift. InternationalIn the United Kingdom, the series premiered on BBC1 on 26 April 1980, in a primetime Saturday night slot. Season 2 began a year later on 8 May 1981, now in a Friday night slot. Season 3 was not shown until late 1983, and then only half of it was shown, at which point the BBC pulled the series from its Friday night slot with no immediate plans of showing any more episodes. The BBC then brought the series back in October 1986, picking up from the middle of season 3 where they last left off, but now it was screened in the afternoon as part of their new daytime line-up. They continued to screen the series until the end, though UK audiences tended to be some 3 – 4 years behind the US. The series concluded in 1996. In France, the show was known as "Côte Ouest" (translated as "West Coast") and was firstly shown on TF1 in 1988 with a new lyrical theme song (in French - two different male singers - several versions). From 2000, the show was rerun from beginning to end on FoxLife, a now defunct satellite channel broadcasting on CanalSat provider.In Germany, the show was known as "Unter der Sonne Kaliforniens" (translated as "Under The Californian Sun"). The premium digital channel Passion currently airs reruns of all episodes.In the Philippines, the show was formerly aired on GMA 7. In Sweden, the show was called "JRs bror – Gary Ewing" (JR's brother – Gary Ewing). The series premiered in 1988 on TV3 (Sweden)TV3. In Ireland, the show was not broadcast on terrestrial TV until 1989 (although viewers with access to BBC were able to watch earlier transmissions). RTE first broadcast the series daily in late 1989 as part of its early afternoon schedule. It proved to be popular and was eventually given an early evening timeslot in February 1991 when RTE moved the show (after showing the first nine seasons) from a daily afternoon timeslot, to a weekly primetime Thursday night slot on its sister channel Network Two (now RTÉ Two). All remaining episodes were shown without a break (the end of each season was followed by the first episode of the next season the following week) until Christmas 1993 when the final episode was broadcast followed by the retrospective Knots Landing Block Party.In Israel, The show aired on The Family Channel (later on renamed as Channel 3) on Cable TV on Fridays evenings at 19:00 from 1990 for the entire first four seasons. Later on, it aired on Sundays at 21:45 for the season & a half that followed, from February 1992. In November 1992, the show was rerun from the beginning, every weekday evening at 20:45. All 14 seasons had aired by April 1994, making it the most popular foreign drama series at this run. From late 1995, the show was rerun in its entirety again in the afternoons (15:50) and the last five seasons at night (01:30). Back To The Cul-De-Sac aired as a holiday special in Autumn 1997.In Italy, the first season was named Da Dallas a Knots Landing (translated: "From Dallas to Knots Landing") though was eventually rebroadcast with the definitive name of "California". Only nine seasons were aired.In Spain, the series was aired by the Regional channels under the FORTA umbrella. The series did not arrive in Spain until at least 1992, and initially was only aired in various regions, normally in a weekday afternoon slot, accompanying Dallas which was greatly delayed in its Spanish TV screenings. Spanish TV never completed the entire run for Knots Landing, nor did it ever complete the Dallas run. Neither series have been rescreened in Spain in full or in part. External links Knots Landing at the Internet Movie Database Knots Landing at TV.com KnotsLanding.Net Official Guide to the Series SOAPnet KL Page 2005 cast reunion photo

The Dean Martin Celebrity Roast

WARNING: Article could not be rendered - ouputting plain text.
Potential causes of the problem are: (a) a bug in the pdf-writer software (b) problematic Mediawiki markup (c) table is too wide

The Dean Martin Celebrity RoastGenreVariety showVariety/Television comedyComedyWritten by David Axelroad, Bill DaleyDirected byGreg GarrisonPresented byDean MartinCountry of origin United StatesLanguage(s) EnglishNo. of seasons 10No. of episodes 54: 29 Specials, 25 as Dean Martin Show segmentProductionProducer(s) Greg GarrisonRunning time VariesBroadcastOriginal channelNBCOriginal run October 31, 1974 – December 7, 1984ChronologyPreceded byThe Dean Martin Show The Dean Martin Celebrity Roast is a NBC television special show hosted by entertainer Dean Martin from 1974 to 1984. For a series of 54 specials and shows, Martin would periodically "Roast (comedy)roast" a celebrity. These roasts were patterned after the roasts held at the New York Friars' Club in New York City. The format would have the celebrity guest seated at a banquet table, and one by one the guest of honor was affectionately chided or insulted about his career by his fellow celebrity friends.In 1973, The Dean Martin Show was declining in popularity. The final (1973–74) season of his variety show would be retooled into one of celebrity roasts, requiring less of Martin's involvement. For the 1973-1974 season, a new feature called "Man of the Week Celebrity Roast" was added to try to pick up the ratings. The roasts seemed to be popular among television audiences and are often marketed in post-issues as part of the official Dean Martin Celebrity Roasts and not The Dean Martin Show. After The Dean Martin Show was cancelled in 1974, NBC drew up a contract with Martin to do several specials and do more roast specials. Starting with Bob Hope in 1974, the roast was taped in California and turned out to be a hit, leading to many other roasts to follow.Memories Are Made of This: Dean Martin Through His Daughter's Eyes. Deana Martin, Wendy Holden Random House Digital, Inc., 2005 ISBN 1-4000-9833-5In the fall of 1974, the roasts moved permanently to the Bally's Las VegasMGM Grand Hotel's Ziegfeld Room in Las Vegas ValleyLas Vegas and mainly aired Thursdays on NBC. The televised roasts were popular in the ratings; however Martin and NBC declined to extend the 10 year contract. Some segments were taped prior or after the roast, due to considerations with the performer or technical aspects.The Rat Pack: neon nights with the kings of cool. Lawrence J. Quirk, William Schoell. HarperCollins, 1999. ISBN 0-380-73222-X Martin hosted no roasts from 1980 through 1983. The show's official title as a television special would change based on the celebrity, in James StewartJimmy Stewart's case for instance, it would be the Dean Martin Celebrity Roast: Jimmy Stewart.The "Roastees" were also referred to as "Man of the Hour" or "Woman of the Hour" ("Man of the Week" in earlier episodes). Martin roasted two sets of celebrities, Jack Klugman and Tony Randall, and Dan Rowan and Dick Martin (comedian)Dick Martin. Only one person was honored posthumously, George Washington, who was honored for the upcoming United States bicentennial (veteran historical impersonator Jan Leighton portrayed Washington for the episode while Audrey Meadows portrayed his wife, Martha WashingtonMartha). Michael Landon, Redd Foxx, Joe Namath, and Jack Klugman were the only celebrities roasted twice; Landon's second time in 1984, being the final roast. For Dean Martin's roast, Don Rickles was host, and assumed the role of Roastmaster.Historical dictionary of African-American television.Volume 7 of Historical dictionaries of literature and the arts. Kathleen Fearn-Banks, Scarecrow Press, 2006. ISBN 0-8108-5335-3 Comedian Nipsey Russell and impressionist Rich Little appeared the most often on the roast with each appearing 24 times. The Dean Martin Celebrity Roasts were released on DVD through Guthy-Renker with some of The Dean Martin Show roasts selected as part of the package. The show is one of the most sold video sets of all-time.Memories Are Made of This: Dean Martin Through His Daughter's Eyes. Deana Martin, Wendy Holden Random House Digital, Inc., 2005 ISBN 1-4000-9833-5 NBC Universal brought suit against producer Greg Garrison and Guthy-Renker for selling The Dean Martin Show DVDs; the suit did not affect the Celebrity Roasts.Roastees, Air Date and RoastersAs The Dean Martin ShowRonald Reagan, 9/13/1973 Roasters: Jackie Vernon (comedian)Jackie Vernon, Phyllis Diller, Dom DeLuise, Pat Henry (comedian)Pat Henry, Mark Spitz, Jack Benny, Jonathan Winters, Nipsey Russell, Nancy Reagan, Don RicklesHugh Hefner, 9/20/1973 Roasters: Billy Baxter (comedian)Billy Baxter, Howard Cosell, Jackie Gayle, Dick Martin (comedian)Dick Martin, Joey BishopEd McMahon, 9/27/1973 Roasters: Pat Buttram, Jackie Vernon (comedian)Jackie Vernon, Dionne Warwick, Steve Landesberg, Charo, Jack Carter (comedian)Jack CarterWilliam Conrad, 10/4/1973 Roasters: Phyllis Diller, Nipsey Russell, Petula Clark, Bob Newhart, Jackie GayleKirk Douglas, 10/11/1973 Roasters: Ted Knight, Norm Crosby, Jack Burns, Avery Schreiber, Rich Little, Lynn Anderson, Jackie Gayle, Don RicklesBette Davis, 10/18/1973 Roasters: Pat Buttram, Kay Medford, Nipsey Russell, Joyce Haber, Vincent Price, Army Archerd, Barbara Heller, Henry Fonda, Howard CosellBarry Goldwater, 10/25/1973 Roasters: William Conrad, Norm Crosby, Steve Landesberg, Carroll O'Connor, William Holden, Mark Russell, Zsa Zsa Gabor, Donald Rice, Dan RowanJohnny Carson, 11/1/1973 Roasters: George Burns, Truman Capote, Doc Severinsen, Joey Bishop, Ruth Buzzi, Dom DeLuise, Bob Newhart, Louisa Moritz, Frederick de CordovaFred DeCordova, Jonathan Winters, Foster Brooks, Dionne Warwick, Rich Little, Barry Goldwater, Bette Davis, Martin Milner, Kent McCord, Redd Foxx, Jack Benny, Joanna HollandWilt Chamberlain, 11/8/1973 Roasters: Norm Crosby, Happy Hairston, Audrey Meadows, Ken Berry, Nipsey Russell, Bill Shoemaker, George Kennedy, Jackie GayleHubert Humphrey, 11/22/1973 Roasters: Pat Henry (comedian)Pat Henry, Nipsey Russell, Gene Kelly, Leo Durocher, Mort Sahl, Ted Knight, Mark Russell, Rich Little, Donald Rice, Lowell P. Weicker, Jr.Lowell Weicker, Foster Brooks, Audrey MeadowsCarroll O'Connor, 12/7/1973 Roasters: Dan Rowan, Dick Martin (comedian)Dick Martin, Marty Allen (comedian)Marty Allen, Mike Connors, Elsie Birdwell, Norm Crosby, John Lindsay, Nipsey Russell, Gene Kelly, Joey Bishop, Ward Wood, William Conrad, William Holden, Donald O'Connor, Charles Nelson Reilly, Zsa Zsa Gabor, Ruth Buzzi, Barry Goldwater, Cass Elliot, Redd Foxx, Don Rickles, Foster BrooksMonty Hall, 12/13/1973 Roasters: Bert Parks, Audrey Meadows, Rocky Graziano, Donald O'Connor, Gene Kelly, Birch Bayh, Art Linkletter, Jack Carter (comedian)Jack CarterJack Klugman & Tony Randall, 12/20/1973 Roasters: Mike Connors, Gary Burghoff, Ruth Buzzi, Soupy Sales, Jack Carter (comedian)Jack Carter, Leonard Barr, Jackie Vernon (comedian)Jackie Vernon, Loretta Lynn, Doug Kershaw, Laffit Pincay, Jr.Zsa Zsa Gabor, 1/10/1974 Roasters: Donald O'Connor, Charles Nelson Reilly, Sue Cameron, Corbett Monica, Donna Fargo, Lonnie Schorr, Ruth Buzzi, Buddy Hackett, Mel TillisLeo Durocher, 1/17/1974 Roasters: Gene Kelly, Maury Wills, Bobby Riggs, Dizzy Dean, Chuck Connors, Alex Karras, Jack Carter (comedian)Jack Carter, Foster Brooks, Gladys KnightTruman Capote, 1/24/1974 Roasters: Jean Simmons, Ted Knight, Audrey Meadows, Donald O'Connor, Rich Little, Joseph Wambaugh, Rocky Graziano, Foster Brooks, Anna Moffo, Johnny RussellDon Rickles, 2/7/1974 Roasters: Joey Bishop, Phyllis Diller, Lorne Greene, Dan Rowan, Dick Martin (comedian)Dick Martin, Casey Kasem, Bob Newhart, Carol Channing, Nipsey Russell, Cliff Robertson, Jack Klugman, Pat Henry (comedian)Pat Henry, Kirk Douglas, Rich Little, Telly Savalas, Charlie Callas, Foster Brooks, Eugene CernanRalph Nader, 2/14/1974 Roasters: Jack Klugman, James Brolin, Jane Withers, Steve Landesberg, Mort Sahl, Rich Little, Ruth Buzzi, Jackie Gayle, Foster Brooks, Donna FargoJack Benny, 2/21/1974 Roasters: James StewartJimmy Stewart, Joey Bishop, Florence Henderson, George Burns, Norm Crosby, Zubin Mehta, Pearl Bailey, Dick Martin (comedian)Dick Martin, Mark Spitz, Wayne Newton, Rich Little, Demond Wilson, Jack Carter (comedian)Jack Carter, Foster Brooks, Gary BurghoffRedd Foxx, 3/1/1974 Roasters: Demond Wilson, Joey Bishop, Nipsey Russell, Foster Brooks, Norm Crosby, Rich Little, Jim Bailey (entertainer)Jim Bailey, Jackie Gayle, John Barbour (actor)John Barbour, Slappy WhiteBobby Riggs, 3/8/1974 Roasters: Vincent Price, Chuck Connors, Leo Durocher, Alex Karras, Wayne Newton, Lynn Anderson, Don Rice, Jack Carter (comedian)Jack Carter, Rosemary Casals. George Washington (portrayed by Jan Leighton, 3/15/1974 Audrey Meadows (as Martha Washington), Dick Martin (comedian)Dick Martin, Leonard Barr, Steve Lawrence (as Washington's aide-de-camp), Corbett Monica, Nipsey Russell, Euell Gibbons, Henny Youngman, Jack Carter (comedian)Jack Carter, Foster Brooks (as Washington's press secretary) Dan Rowan & Dick Martin (comedian)Dick Martin, 3/22/1974 Roasters: Joey Bishop, Audrey Meadows, Arte Johnson, Richard Dawson, Nipsey Russell, Steve

Lawrence, Bob Newhart, Ruth Buzzi, Foster BrooksHank Aaron, 3/29/1974 Roasters: Joey Bishop, Eddie Mathews, Audrey Meadows, Lou Rawls, Norm Crosby, Jackie Bavene, Lynn Anderson, Nipsey Russell, Dizzy Dean, Jeannie Rineal, Rodney Allen Rippy, Foster BrooksJoe Namath, 4/5/1974 Roasters: Don Meredith, Weeb Ewbank, Bear BryantPaul "Bear" Bryant, Dick Butkus, Jim Plunkett, David Janssen, Angie Dickinson, Fulton J. Sheen, Foster Brooks, Charlie Callas, Joey Bishop, Rich Little, Slappy WhiteAs The Dean Martin Celebrity RoastBob Hope, 10/31/1974 Roasters: Flip Wilson, Ronald Reagan, John Wayne, Ginger Rogers, Foster Brooks, James StewartJimmy Stewart, Milton Berle, Billy Graham, Rich Little, Howard Cosell, Jack Benny, Zsa Zsa Gabor, Nipsey Russell, Omar Bradley, Phyllis Diller, Neil Armstrong, Henry Kissinger, Don Rickles, Dolores HopeTelly Savalas, 11/15/1974 Roasters: Phyllis Diller, Howard Cosell, George Kennedy, Dom DeLuise, Peter Graves, Casey Kasem, Shelley Winters, Ernest Borgnine, Dan Rowan, Dick Martin (comedian)Dick Martin, Foster Brooks, Richard Roundtree, Robert Stack, Nipsey Russell, Rich Little, Angie Dickinson, Don RicklesLucille Ball, 2/7/1975 Roasters: Phyllis Diller, Dan Rowan, Dick Martin (comedian)Dick Martin, Ruth Buzzi, Bob Hope, Milton Berle, Gary Morton, Gale Gordon, Totie Fields, Rich Little, Henry Fonda, Ginger Rogers, Foster Brooks, Nipsey Russell, Jack Benny, Vivian Vance, Don RicklesJackie Gleason, 2/27/1975 Roasters: Phyllis Diller, Milton Berle, Danny Thomas, Audrey Meadows, Gene Kelly, Nipsey Russell, Sandy Hirt, Sid Caesar, Sheila MacRae, Art Carney, Frank Gorshin, Foster BrooksSammy Davis, Jr., 4/24/1975 Roasters: Milton Berle, Wilt Chamberlain, Freddie Prinze, Norm Crosby, Dionne Warwick, Joey Bishop, Nipsey Russell, Phyllis Diller, Jan Murray, Frank Gorshin, Foster Brooks, Don Rickles, Altovise DavisMichael Landon, 5/15/1975 Roaster: Amanda Blake, Ernest Borgnine, Sid Caesar, Phyllis Diller, Euell Gibbons, Lorne Greene, Guy Marks, Jan Murray, Don Rickles, Cliff Robertson, Victor Sen Yung, Joey Bishop, Norm CrosbyEvel Knievel, 11/10/1975 Roasters: Gabe Kaplan, Isabel Sanford, Ernest Borgnine, Milton Berle, William Conrad, Barry Goldwater, Glen Campbell, Georgia Engel, Jackie Cooper, Nipsey Russell, Joyce Brothers, Charlie Callas, Cliff Robertson, McLean Stevenson, Audrey Meadows, Ruth Buzzi, Don RicklesValerie Harper, 11/20/1975 Roasters: Rich Little, Isabel Sanford, David Groh, Red Buttons, Phyllis Diller, Jack Albertson, Shelley Winters, Milton Berle, Edward AsnerEd Asner, Richard Schaal, Harold Gould, Eva Gabor, Chad Everett, Julie Kavner, Nipsey Russell, Georgia Engel, Foster Brooks, Jack Carter (comedian)Jack Carter. Muhammad Ali, 2/19/1976 Roasters: Freddie Prinze, Gabe Kaplan, Red Buttons, Tony Orlando, Wilt Chamberlain, Georgia Engel, Billy Crystal, Floyd Patterson, Howard Cosell, Ruth Buzzi, Gene Kelly, Sherman Hemsley, Isabel Sanford, Rocky Graziano, Foster Brooks, Orson Welles, Nipsey RussellDean Martin, 2/27/1976 Roasters: Don Rickles (Roastmaster), Orson Welles, Paul Lynde, Joe Namath, Barry Goldwater, Angie Dickinson, Muhammad Ali, James StewartJimmy Stewart, Gabe Kaplan, Gene Kelly, Hubert Humphrey, Charlie Callas, John Wayne, Joey Bishop, Rich Little, Ruth Buzzi, Tony Orlando, Georgia Engel, Nipsey Russell, Foster Brooks, Howard Cosell, Dan Rowan, Dick Martin (comedian)Dick Martin, Bob HopeDennis Weaver, 4/27/1976 Roasters: William Conrad, Shelley Winters, Steve Forrest (actor)Steve Forrest, Red Buttons, Rich Little, Mike Connors, Milburn Stone, Ruth Buzzi, Nipsey Russell, Zsa Zsa Gabor, Georgia Engel, Amanda Blake, Peter Graves, Foster Brooks, Milton BerleJoe Garagiola, Sr.Joe Garagiola, 5/25/1976 Roasters: Red Buttons, Orson Welles, Luis Tiant, Pat Henry (comedian)Pat Henry, Nipsey Russell, Willie Mays, Hank Aaron, Yogi Berra, Norm Crosby, Shirley Jones, Mickey Mantle, Jack Carter (comedian)Jack Carter, Stan Musial, Maury Wills, Charlie FinleyCharles O. Finley, Charlie Callas, Jackie GayleRedd Foxx, 11/26/1976 Roasters: Demond Wilson, Milton Berle, Isabel Sanford, Isaac Hayes, LaWanda Page, Jimmie Walker, Slappy White, Liz Torres, George Kirby, Norm Crosby, Steve Allen, Joe Garagiola, Sr.Joe Garagiola, Nipsey Russell, Don Rickles, Marty Allen (comedian)Marty Allen, Abe VigodaDanny Thomas, 12/15/1976 Roasters: Lucille Ball, Milton Berle, Gene Kelly, Jimmie Walker, Red Buttons, Harvey Korman, Charo, Charlie Callas, Ruth Buzzi, Nipsey Russell, Dena Dietrich, Howard Cosell, Jan Murray, Don Knotts, Orson WellesAngie Dickinson, 2/8/1977 Roasters: James StewartJimmy Stewart, Eve Arden, Earl Holliman, Orson Welles, Joey Bishop, Juliet Prowse, Jimmie Walker, Cindy Williams, Red Buttons, Cathy Rigby, Jackie Mason, Scatman Crothers, LaWanda Page, Ruth Buzzi, Charlie Callas, Foster Brooks, Rex ReedGabe Kaplan, 2/21/1977 Roasters: Nipsey Russell, Liz Torres, Billy Crystal, Red Buttons, Charo, Jimmie Walker, Orson Welles, Howard Cosell, Johnny Bench, Ed Bluestone, Alice Ghostley, Joe Garagiola, Sr.Joe Garagiola, Charlie Callas, Abe Vigoda, George

Kirby, Milton BerleTed Knight, 3/2/1977 Roasters: Edward AsnerEd Asner, Gavin MacLeod, Georgia Engel, James StewartJimmy Stewart, Orson Welles, Jimmie Walker, Jack Carter (comedian)Jack Carter, Harvey Korman, Jackie Wilson, Red Buttons, LaWanda Page, Paul Williams (songwriter)Paul Williams, Foster Brooks, Scatman Crothers, Kelly Monteith, Willy Tyler & Lester, Julie McWhirter, Renée RichardsPeter Marshall (U.S. entertainer)Peter Marshall, 5/22/1977 Roasters: Joey Bishop, Rose Marie, Red Buttons, Zsa Zsa Gabor, Orson Welles, Rip Taylor, Vincent Price, Karen Valentine, Ed Bluestone, Foster Brooks, Jimmie Walker, Jackie Gayle, Paul Lynde, Jack Carter (comedian)Jack Carter, Wayland Flowers & Madam Dan Haggerty, 11/2/1977 Roasters: Denver Pyle, William Conrad, Harry Morgan, Pat Harrington, Jr.Pat Harrington, Roger Miller, Orson Welles, Abe Vigoda, Rich Little, Ruth Buzzi, Red Buttons, Jimmie Walker, Charlie Callas, LaWanda Page, Foster Brooks, Marilyn Michaels, Jackie Gayle, Tom DreesenFrank Sinatra, 2/7/1978 Roasters: Ronald Reagan, Gene Kelly, Don Rickles, George Burns, Dom DeLuise, Redd Foxx, James StewartJimmy Stewart, Flip Wilson, Telly Savalas, Jonathan Winters, LaWanda Page, Red Buttons, Milton Berle, Ernest Borgnine, Orson Welles, Rich Little, Barbara SinatraBarbara Marx, Jilly Rizzo, Ruth Buzzi, Peter FalkJack Klugman, 3/17/1978 Roasters: Tony Randall, Milton Berle, Katherine Helmond, Dick Martin (comedian)Dick Martin, Red Buttons, Abbe Lane, Joey Bishop, Phyllis Diller, LaWanda Page, Ruth Buzzi, Robert Guillaume, Kay Medford, Connie Stevens, Foster Brooks, Don Rickles, Joyce BrothersJames StewartJimmy Stewart, 5/10/1978 Roasters: Lucille Ball, June Allyson, Greer Garson, Red Buttons, Barry Goldwater, Henry Fonda, Eddie Albert, Foster Brooks, George Burns, Tony Randall, Don Rickles, Rich Little, Janet Leigh, Milton Berle, Jesse White (actor)Jesse White, Orson Welles, Mickey Rooney, LaWanda Page, Ruth BuzziGeorge Burns, 5/17/1978 Roasters: Gene Kelly, Connie Stevens, James Stewart, Milton Berle, Jack Carter (comedian)Jack Carter, Phyllis Diller, Ernest Borgnine, Joey Bishop, Norm Crosby, Euell Gibbons, Guy Marks, Jan Murray, Cliff Robertson, Lorne Greene, Sid Caesar, Amanda Blake, Victor Sen Yung, Don Rickles, LaWanda Page, Abe Vigoda, Dom DeLuise, Ronald Reagan, Orson Welles, Charlie Callas, Ruth BuzziBetty White, 5/31/1978 Roasters: Peter Marshall (U.S. entertainer)Peter Marshall, Jimmie Walker, Bonnie Franklin, Orson Welles, Red Buttons, Charlie Callas, Abe Vigoda, LaWanda Page, Foster Brooks, Phyllis Diller, Rich Little, John Hillerman, Georgia Engel, Dan Haggerty, Milton Berle, Allen LuddenSuzanne Somers, 11/21/1978 Roasters: Milton Berle, Lorne Greene, Joyce Brothers, Audra Lindley, Norman Fell, Norm Crosby, Lee Meriwether, Charlie Callas, Bernie Kopell, Red Buttons, LaWanda Page, Paul Anka, Ruth Buzzi, Tom Bosley, Rich Little, Jackie Gayle, Zsa Zsa Gabor, Alan HamelJoe Namath, 1/19/1979 Roasters: Joey Bishop, Bear BryantPaul "Bear" Bryant, Slappy White, Dick Butkus, Corbett Monica, Angie Dickinson, Audrey Meadows, Charlie Callas, David Janssen, Nipsey Russell, Jim Plunkett, Don Meredith, Rich Little, Fulton J. Sheen, Jack Carter (comedian)Jack Carter, Weeb Ewbank, Foster Brooks, Mel Tillis, Bernie Kopell, David Doyle (actor)David Doyle, George BlandaJoan Collins, 2/24/1984 Roasters: John Forsythe, Aaron Spelling, Gavin MacLeod, Bea ArthurBeatrice Arthur, Red Buttons, Phyllis Diller, Angie Dickinson, Charlie Callas, Rich Little, Anne Baxter, Don Rickles, Zsa Zsa Gabor, Dom DeLuise, Milton BerleMr. T, 3/14/1984 Roasters: Bob Hope, George Peppard, Gary Coleman, Ann Jillian, Red Buttons, Rich Little, Rick Schroder, Gavin MacLeod, Slappy White, Dick Shawn, Howard Cosell, Maureen Murphy (comedian)Maureen Murphy, Don Rickles, Nell CarterMichael Landon, 12/7/1984 Roasters: Melissa Gilbert, Brian Keith, Lorne Greene, Merlin Olsen, Vic Tayback, Maureen Murphy (comedian)Maureen Murphy, Rich Little, Pat Harrington, Jr.Pat Harrington, Dick Butkus, Bubba Smith, Dick Shawn, Norm Crosby, Orson Welles, Don Rickles, Slappy WhiteSee alsoComedy Central RoastReferences

Century_Village,_Florida

Century Village, Florida	
— CDP —	
Coordinates: 26°42′56″N 80°7′34″W	
Country	United States
State	Florida
County	Palm Beach
Area	
• **Total**	1.1 sq mi (2.8 km^2)
• **Land**	1.0 sq mi (2.6 km^2)
• **Water**	0.1 sq mi (0.2 km^2)
Elevation	16 ft (5 m)
Population (2000)	
• **Total**	7616
• **Density**	7588.1/sq mi (2929.8/km^2)
Time zone	Eastern (EST) (UTC-5)
• **Summer (DST)**	EDT (UTC-4)
FIPS code	12-11437[1]
GNIS feature ID	1867125[2]

Century Village is a census-designated place (CDP) in Palm Beach County, Florida,[3] United States (west of the City of West Palm Beach) and is best known as a retirement community.

There are three other Century Village developments in Broward and Palm Beach Counties (in Deerfield Beach, Boca Raton and Pembroke Pines) which are not part of the Century Village CDP.

Demographics

As of the census[1] of 2000, there were 7,616 people, 5,536 households, and 1,777 families residing in the CDP. The population density was 2,940.6/km² (7,588.1/mi²). There were 8,754 housing units at an average density of 3,379.9/km² (8,721.9/mi²). The racial makeup of the CDP was 98.41% White (95.6% were Non-Hispanic White,)[4] 0.43% African American, 0.03% Native American, 0.42% Asian, 0.08% Pacific Islander, 0.11% from other races, and 0.53% from two or more races. Hispanic or Latino of any race were 2.99% of the population.

There were 5,536 households out of which 0.1% had children under the age of 18 living with them, 29.0% were married couples living together, 2.4% had a female householder with no husband present, and 67.9% were non-families. 65.2% of all households were made up of individuals and 57.4% had someone living alone who was 65 years of age or older. The average household size was 1.38 and the average family size was 2.07.

In the CDP the population was spread out with 0.3% under the age of 18, 0.3% from 18 to 24, 1.9% from 25 to 44, 12.9% from 45 to 64, and 84.6% who were 65 years of age or older. The median age was 78 years. For every 100 females there were 61.1 males. For every 100 females age 18 and over, there were 61.0 males.

The median income for a household in the CDP was $18,780, and the median income for a family was $29,604. Males had a median income of $22,557 versus $22,292 for females. The per capita income for the CDP was $18,920. About 7.2% of families and 12.3% of the population were below the poverty line, including none of those under age 18 and 11.6% of those age 65 or over.

As of 2000, English was the first language for 86.46% of all residents, while Yiddish accounted for 3.77%, Spanish for 3.43%, German was spoken by 1.75%, French made up 1.27%, Russian was totaled at 1.07%, Hebrew totaled 0.74%, Italian was at 0.67%, Polish at 0.47%, and Hungarian was the mother tongue 0.33% of the population.[5]

Century Village in popular culture

- Comedian Red Buttons was a frequent advertising spokesman for Century Village.[6]
- Century Village is sometimes jokingly referred to as "Cemetery Village" due to the high proportion of elderly residents.[7]
- Century Village provided subject matter for a comedy skit on the Phil Hendrie Show on the radio station 610 WIOD.[8]
- The 1996 children's novel *The View from Saturday* by E. L. Konigsburg is partly set in Century Village.

References

[1] "American FactFinder" (http://factfinder.census.gov). United States Census Bureau. . Retrieved 2008-01-31.
[2] "US Board on Geographic Names" (http://geonames.usgs.gov). United States Geological Survey. 2007-10-25. . Retrieved 2008-01-31.
[3] Florida -- Place GCT-PH1. Population, Housing Units, Area, and Density: 2000 (http://factfinder.census.gov/servlet/GCTTable?_bm=y&-geo_id=04000US12&-_box_head_nbr=GCT-PH1&-ds_name=DEC_2000_SF1_U&-format=ST-7) Accessed December 5, 2007
[4] "Demographics of Century Village, Fla." (http://www.muninetguide.com/states/florida/municipality/Century_Village.php). MuniNetGuide.com. . Retrieved 2007-11-21.
[5] "MLA Data Center Results for Century Village, Florida" (http://www.mla.org/map_data_results&state_id=12&county_id=&mode=place&zip=&place_id=11437&cty_id=&ll=&a=&ea=&order=r). Modern Language Association. . Retrieved 2007-12-06.
[6] "Celebrity Tie-ins", Big Builder Magazine, May 1, 2004 (http://www.bigbuilderonline.com/Industry-news.asp?channelID=55&sectionID=27&articleID=55204)
[7] "The new voices of Century Village", Miami Herald, September 16, 2007 (http://www.miamiherald.com/living/home_design/story/237437.html)
[8] PhilHendrieShow.com - Category Clips (http://www.philhendrieshow.com/Radio/Categories.aspx?action=view&id=5&name=Pre+Los+Angeles)

External links

- Century Village website (http://www.centuryvillage.com/)

Anna_Kashfi

Anna Kashfi	
Born	Johanna O'CallaghanSeptember 30, 1934Darjeeling, British India
Occupation	Actress
Years active	1956-1963
Spouse(s)	Marlon Brando (1957-1959) James Hannaford (1974-19??; his death)
Website	
IMDb biodata [1]	

Anna Kashfi (born September 30, 1934) is an Anglo-Indian former film actress, who had a brief Hollywood career in the 1950s and is best known for having been married to Marlon Brando.

Early life and origins

Johanna O'Callaghan was born in Darjeeling and raised in Calcutta, India and, later, Cardiff, Wales, where the family moved in 1947.[1] William P. O'Callaghan was a London-born steel worker of Irish descent, who had been a traffic superintendent on the Indian State railways; his wife was English and told the press *"there is no Indian blood in my family or my husband's family"*.[1] In her book, *Brando for Breakfast*, published in 1979, however, Kashfi claimed that she really is half-Indian and that the press incorrectly believed that William O'Callaghan was her real father when he was, she stated, her stepfather. Kashfi wrote that her biological father was Indian and that she was the result of an "unregistered alliance" between her parents. Film director Edward Dmytryk, who directed the actress in her first film, stated that he knew her real surname was Irish but he said he assumed that she was half-Indian.[1]

Career

She worked as a waitress in Cardiff and in a butcher shop before moving to London, where she became a model[2]

The actress made her debut screen appearance using the stage name *Anna Kashfi* as a Hindu girl in *The Mountain* (1956) for Paramount with Spencer Tracy and Robert Wagner. In her next film she co-starred with Rock Hudson as a Korean girl in *Battle Hymn* (1957). This was followed by *Cowboy* (1958) with Glenn Ford and Jack Lemmon in which she played a Mexican. Her next and last film during this period was *Night of the Quarter Moon* (1959). She made a few appearances on television, including the series *Adventures in Paradise*, though drug and alcohol problems reportedly contributed to the premature end of her acting career[3]

Personal life

Kashfi married Marlon Brando on October 11, 1957, and they were divorced on April 22, 1959. She had met Brando in the summer of 1956 and dated all through that year and all through 1957. According to *The New York Times* (October 14, 1957), though Kashfi's parents had been identified as the O'Callaghans, she did not identify them as her parents on her marriage license, instead stating that her father was Devi Kashfi and her mother was Selma Ghose; a friend of the bride stated that Kashfi's purported Indian father had died six weeks before the Brandos' wedding.[4] The O'Callaghans' were adamant that Kashfi was their child, and William O'Callaghan was quoted in *Time* magazine as saying, "That's our daughter, and both me and missus were born in London." The magazine further stated that *"MGM, which likes Johanna-Anna in her off-shoulder sari, first hedged, then admitted her identity."*[5]

The Brandos had a son, Christian Devi Brando, (1958–2008), and fought bitterly over custody of the boy, with Brando eventually obtaining custody. Kashfi married James Hannaford, a salesman, in 1974.

In the 1990s, her son, Christian, whom she called "Devi", was tried for killing his half-sister Cheyenne Brando's boyfriend, Dag Drollet. Christian Brando died of pneumonia in Los Angeles in 2008, aged 49. Kashfi has one grandson, Michael Brando (born 1988).

Books

- Anna K. Brando and E. P. Stein, *Brando for Breakfast*, Berkley Pub Group, 1980, ISBN 0-425-04698-2.

Filmography

- *The Mountain* (1956)
- *Battle Hymn* (1957)
- *Cowboy* (1958)
- *Night of the Quarter Moon* (1959)

Television appearances

- *Adventures in Paradise* (1959)
- *The Deputy* (1960)
- *Bronco* (1960)

References

- Peter Manso, *Brando. The Biography*, Hyperion, New York, 1994, ISBN 0-7868-6063-4

Notes

[1] "Kashfi Called Welsh", *The New York Times*, 13 October 1957

[2] Anna Kashfi (band) Interview by John Clarkson, September 24, 2004 (http://www.pennyblackmusic.co.uk/MagSitePages/Article.aspx?id=3441), *Brando Unzipped*, Darwin Porter, 2006

[3] *Los Angeles Times* report on Anna Kashfi (http://www.latimes.com/la-me-brando27jan27,1,887238.story)

[4] "Kashfi Still Enigma: License Does Not List Welsh Couple As Parents", *The New York Times*, October 14, 1957

[5] *Time* magazine (http://www.time.com/time/magazine/article/0,9171,937947-1,00.html)

External links

- Anna Kashfi (http://www.imdb.com/name/nm0440542/) at the Internet Movie Database
- Murder by Kashfi's son (http://www.crimelibrary.com/notorious_murders/family/christian_brando/2.html)
- Anna Kashfi bio (http://homepage.ntlworld.com/elizabeth.ercocklly/kashfi.htm)
- Anna Kashfi 'cried tears of rage' upon hearing of Marlon Brando's death (http://archives.tcm.ie/breakingnews/2004/07/11/story156558.asp)

Movita_Castaneda

Movita Castaneda	
 from the trailer for the film *Mutiny on the Bounty* (1935).	
Born	Maria CastenedaDecember 4, 1917Nogales, Arizona, U.S.
Other names	Movita
Years active	1930–1989
Spouse(s)	Marlon Brando (1960–1962) (divorced) 2 children Jack Doyle (1939-1944) (divorced)

Maria "Movita" Castaneda (born December 4, 1917) is an American actress best known for being the second wife of actor Marlon Brando. She was six years older than Brando. In films, she played exotic women/singers, such as in *Flying Down to Rio* (1933) and *Mutiny on the Bounty* (1935), of which she is the last surviving cast member. She is the mother of Miko Castaneda Brando (b. 1961) and Rebecca Brando Kotlizky (b. 1966).

Life and work

Movita, a Mexican American, was born in Nogales, Arizona, on a train travelling between Mexico and Arizona. Movita began her acting career singing the Carioca to Ginger Rogers and Fred Astaire's first dance number in the first film in which the famous duo appeared together, *Flying Down to Rio* (1933). She continued playing exotic women in American and Spanish language films in the 1930s, most notably as a Tahitian girl, Tehanni in *Mutiny on the Bounty* (1935) alongside Clark Gable and Franchot Tone. She played another island girl in *Paradise Isle* (1937), and starred in two films alongside John Carroll, *Rose of the Rio Grande* (1938) and *Wolf Call* (1939), and then in *The Girl from Rio* (1939) with Warren Hull. She starred in the British thriller *Tower of Terror* (1941) alongside Wilfrid Lawson and Michael Rennie. After a break, she appeared as Henry Fonda's cook in *Fort Apache* (1948), then starred with Tim Holt in two further westerns: *The Mysterious Desperado* (1949), and *Saddle Legion* (1951).

In the early 1940s, Movita married the Irish boxer, singer and actor Jack Doyle (known as the Gorgeous Gael). Doyle was noted for disappearing on drinking bouts and lost several important fights while drunk. The marriage did not last long.

After appearing in a few more minor westerns and a few TV parts, she met the actor Marlon Brando in the late 1950s, after his breakup with Anna Kashfi. She had two children with Brando: Miko Brando and Rebecca Brando. Brando played the role of Fletcher Christian in the 1962 remake of the 1935 film in which Movita had played a Tahitian girl, Tehanni. Brando then married his co-star Tarita Teriipia, who played Christian's Tahitian wife Maimiti in the film, with whom he also had two children. After a small TV role in 1977, Movita appeared as Ana in 17 episodes of *Knots Landing* from October 1987 to May 1989.

External links

- Movita Castaneda [1] at the Internet Movie Database

Tarita_Teriipia

Tarita Teriipia	
Born	Tarita TeriipiaDecember 29, 1941Bora Bora, French Polynesia
Years active	1962–1985
Spouse(s)	Marlon Brando (1962–1972) (divorced) 2 children

Tarita Teriipia (born December 29, 1941 in Bora Bora, French Polynesia) is a former actress of French Polynesian and Chinese descent most famous for being the wife of Marlon Brando. She was usually billed as **Tarita**.

Biography

She played Maimiti opposite Marlon Brando in the film *Mutiny on the Bounty* (1962), for which she received a Golden Globe nomination for Best Supporting Actress, and became Brando's third wife in 1962. She is the mother of two of Brando's children, a son, Simon Teihotu, and a daughter, Tarita Cheyenne. They divorced in 1972. Only months after Marlon Brando's death in 2004, Tarita published her memoirs titled *Marlon, My Love and My Torment.* One of their grandsons, Tuki Brando, is a prominent model.

Sources

- Tarita biography [1]
- NY Times about Tarita's memoirs [2]

External links

- Tarita Teriipia [3] at the Internet Movie Database

Article Sources and Contributors

Red_Buttons *Source*: http://en.wikipedia.org/w/index.php?title=Red_Buttons *Contributors*: AHMartin, AdamBMorgan, Akcarver, Alberia torkenkluvin, All Hallow's Wraith, Ameliorate!, Anthropocentrism, Anticipation of a New Lover's Arrival, The, Artemisboy, Asc85, Avipol, Barticus88, Bbsrock, Bdve, Belovedfreak, Ben King, Billy Hathorn, Blaine Washington, Bloomington882005, Bovineboy2008, BuffaloBob, COLA1956, Capitalistroadster, Captain Kirk, Cburnett, Clarityfiend, Cliff1911, CommonsDelinker, Cop 663, D6, DSatz, Dabomb87, DavidLevinson, Dblecros, Dcljr, Dieschwarzetur, Donmike10, Downwards, Edcolins, English1123, EvilSuggestions, Floaterfluss, Foofbun, Gaius Cornelius, Gamaliel, Garion96, GlassCobra, Gr8lyknow, Hahnchen, Hhst, Hmains, Honeylemonz, IZAK, Jack Cox, Jack O'Lantern, Jagarin, Jajhill, Jbmas99, Joefromrandb, Joey80, Jpfagerback, JustAGal, Jwrosenzweig, Jzummak, Kaszeta, Kate, Kumioko (renamed), Leofric1, LightBulbFilm, Lore aura, Martarius, MegX, Michael Drew, Michael Snow, MisfitToys, Mo0, Monkeyzpop, Mordicai, MovieMadness, Moviefan, MrDolomite, Music2611, Nehrams2020, Noirish, NorthernThunder, Oobopshark, Paul MacDermott, Pawl Kennedy, Pegship, PhantomS, Piolinfax, Pissant, Primogen, Qylecoop, RMHED, Red Buttons, Reedmalloy, Rentaferret, Rjwilmsi, Rmhermen, Rossrs, Salamurai, Sam, Shadow2700, ShelfSkewed, Sjdunn9, Sogospelman, Stundra, Sumiyoshi, Thatguyflint, Thismightbezach, Tim riley, Tim!, Tinton5, User At Work, VeryVerily, Volatile, We hope, Who, Wikix, Wildhartlivie, William Avery, Willking1979, Wjhonson, Wmahan, Young Kreisler, Zoe, Zoidbergmd, Zzyzx11, 106 anonymous edits

Academy_Award *Source*: http://en.wikipedia.org/w/index.php?title=Academy_Award *Contributors*: *drew, -kayac71-, 1 Lucky Texan, 205ywmpq, 28bytes, 3finger, 5Q5, 75tpickupsx1983, 7dollarsand2cents, A Clown in the Dark, A wild Rattata, A.K.A.47, ADyuaa, AKR619, ARGOU, Aaron Kashtan, Aaron Schulz, AaronY, Aashaa, Abcdefgy2, Academic Challenger, Accurizer, Adam7davies, AdamJacobMuller, Addshore, Adieu Aschau, Adoniscik, Aewold, Afabbro, AgarwalSumeet, Agateller, Agerskov, Ahmad.ghamdi.24, Aidenkaramanyan, Aillema, Aimeemay, Ajcadoo, Alan Liefting, Alansohn, Alaska97, AldezD, Ale jrb, Alejandro Alvarez, AlexManMaster, Alexf, Alexissenoski, AlexiusHoratius, Alfiboy, Allstarecho, Aloysiushonig, Alsharine, Alton, Amelee1, Ameliorate!, American Eagle, AmericanSights2008, Amitkinger, Amy Larie Hall, Anansi00, Andres, Andrevan, Andy M. Wang, Angela, Animaliv, Anishniranam, Aniten21, Anna Lincoln, AnnaFrance, AnonMoos, Anonymous Dissident, Anshuk, Antonio en las Ruedas, Ap, Aphid360, AppuruPan, Aranel, Arielbelle1621, ArkansasTraveler, Arnav Agrawal, Arsonal, Art LaPella, Art1991, Arthena, Artihcus022, Atub1, AuburnPilot, Avia, Avicennasis, Avoided, Avs5221, Awardwiki, Awiseman, Awostrack, AxelBoldt, Aykantspel, BD2412, BDS2006, BIG4PAPA, BNLfan53, Backslash Forwardslash, Bad Graphics Ghost, Balthazarduju, Barticus88, Batmanand, Bazzargh, Bellwether BC, Ben b, Ben-Zin, Bender235, Benedictine, Benzy19, Betaeleven, Beyondthislife, Bfigura's puppy, Bgoldnyxnet, Bhamv, Bibblesone, BigCow, BigDom, Bigvinu, Billybob923, Bilsonius, Binx, Birdienest81, Bjbeamish, Blahblah32blahblah, Blender34, Blue387, BlueLotusLK, Bluejena, Bluetooth22, Bnlmarshie, BobBBoobsII, Bobblewik, Bobet, Bobrayner, Bogsat, BomBom, Bongwarrior, Bookandcoffee, Bookofjude, BoomerAB, Borgx, Borisblue, Bovineboy2008, Bradley0110, Branddobbe, Bremen, Brianbarney, Brianga, BriscX, Browsee, Brusegadi, Bryan Derksen, BryanD, BryanG, Bsd987, Budget6300, Bueller 007, BuickCenturyDriver, BullRangifer, Byelf2007, C-b90, CHEWBACAJC, CJMylentz, CQJ, Cakeshows, Calabe1992, Calmer Waters, Caltas, Camembert, Can't sleep, clown will eat me, Cantus, Carajou, CardinalDan, Carolineroman, Casper Claiborne, Catpowerzzz, Cayla, Cburnett, Cenarium, Centrx, Cescoby, Cesevern, Cgersten, Chancemill, Chantalt1, Chaojoker, Charlytheron, Chcoc, Chicheley, ChowRiit, Chowbok, ChrisGualtieri, Chrishmt0423, Chrishy man, Chrislk02, Chumba1, CieloEstrellado, Cirt, Civvi, Cjdiamond2010, Ck lostsword, Ckatz, Clashfrankcastle, Classic 971, Cleaningdepartment, Clemwang, Cliff smith, ClockworkSoul, Closedmouth, Cmichael, Cocoberrychip, Coffee, Colfer2, Colonies Chris, Comixanimation, CommonsDelinker, ConradPino, Conscious, Conversion script, CookieMunster, Coolmallu, Cop 663, Corpx, Corvus cornix, Cosmic Latte, Courcelles, Courtens, Cpryd001, Cptmurdok, Crakkerjakk, Crazyberns, Crboyer, Crimson Observer, Cssiitcic, Cubs Fan, Cuervo7, Cullengc, Czj, DARTH SIDIOUS 2, DBigXray, DDima, DJ Clayworth, Dabigkid, Daglaton, Daimengrui, Dale Arnett, Damian Yerrick, Daniel J. Leivick, Danielfolsom, Danny, Dannycas, Darkday, Darkwind, Darrenhusted, Darth Panda, Datclever09, DavidFarmbrough, DavidGC, Davidavidjas, Davodd, Dback, Dcljr, Dcoetzee, Deacon of Pndapetzim, Deaflympic, Decltype, DeltaSceptile, Den fjättrade ankan, Deor, DerHexer, Dgabbard, Dgw, Diaa abdelmoneim, Dickytheorienteer, Dictablanda, Diego Moya, Discospinster, Dismas, Disney avatar, Djmixster, Djw bklyn, Dl2000, Dlabtot, Dmz5, Dobie80, Doc Strange, DocWatson42, Donfbreed, Donmike10, Donovan rayson, Donreed, Dori, Doulos Christos, Dputig07, Dr Aaij, Dr. Blofeld, Dr.K., DrWJK, Drake the man, Drdr1989, Dreadstar, Drm310, Drmargi, Drunkenpeter99, Dufc2, Duggy 1138, Duke53, Duncanbruce, Dvyjones, Dyingontheinside-x, DylanW, Dysepsion, Dysprosia, EVula, Eaglecorpse, EamonnPKeane, Earl CG, Earl Manchester, Easily ruined, EclecticEnnui, Eclecticology, Ed Poor, Editore99, Edlitz36, Edokter, Eduardoferrer2k, Edward, Eeekster, Egmontaz, Egnalebd, Ehwills, Electiontechnology, Eliasemanuelbrandanfranco, Elink ii, Elmmapleoakpine, Elvenscout742, Emerson7, EmmyWinner, Empoor, EncMstr, Enderminh, EnekoR, EntChickie, Entheta, Enumclown, Eranb, Erich Schmidt, Ericoides, Erik Horrell, ErinHowarth, EstendorLin, Esuzu, Evanpierce313, Evanreyes, EvieZamora, Excirial, Exitmoose, F, F2003305, FA Jon, FFlixx7481, Face, Falcon8765, Fastily, Favonian, Fbv65edel, Fdp, Femto, Fieldday-sunday, Finalius, Finalnight, Firefox13, Firien, FishPhileo, Fistful of Questions, Flamurai, Flatbush52-1, Flatterworld, Flernk, Flowanda, Fluri, Forbsey, Forrando, Fourthcourse, Foxmulder, Franciscoh, FredR, FreeKresge, Freiwilliger, Friedo, Funkadillo, Funkamatic, Funnysheep, Funnytoes, FuriousFreddy, Fusionmix, Fuzzyslob, GATEkeeper, GFHandel, Gadfium, Gaia Octavia Agrippa, Gareth E Kegg, Garethfoot, Gargamel38, Gea352, Ged UK, Geomr, George Pelltier, Ghimboueils, Gida0501, Gilgamesh, Gilliam, Glacialfox, Glenn, Glennwells, Goatdog, Godardesque, Gohst, Goldcoast1, Good Olfactory, GraemeL, Grandpafootsoldier, Greg Anthony, Greg Bell, Gregfitzy, Gridzel, GringoInChile, Grolltech, Groucho734, Grubel, Grutness, Guest9999, GuidoGuanz, Gurchzilla, Gurgis4163, Gurubrahma, Gwernol, HJ Mitchell, Hadlock, Haleebop, HamburgerRadio, Hankwang, Hapleworth, HarshAJ, Hathawayc, Haunti, Hawaiian717, Hayden Burley, Haymouse, Hazydan, Headbomb, HeckXX, Heimstern, Hephaestos, Hersheysquirts123, HexaChord, Hiphats, Hiplibrarianship, Hmains, Hollollo, Homalzeebala, Honbicot, Hotlorp, Howeman, Hprager, Hqb, Hraefen, HueSatLum, Hulmem, Hut 8.5, Huwa, Iamvered, IanManka, Ianblair23, IankasLoyalar, Ibanez RYM, Ibbn, IceHunter, Icecool1077, Iced Kola, Idleguy, Ig88b, Ilovekelly75, Ilse@, Ilumina93, InaVegt, Ineverwonanoscar, Ingebjorg, Innotata, Inpenetrableman, Invisigoth65, Irbisgreif, Irishguy, Irishman37, Ironfist06, J Raab, J.delanoy, J.smith, J36miles, JCO312, JD554, JForget, JJIG, JSydel, JackO'Lantern, JackofOz, Jackson65, Jagarin, Jagged 85, Jahangard, Jalabi99, Jamccull, JamesAM, JamesLucas, Janbryan, Janke, Jaret Green, Jasonbres, Jatterb, Java7837, Jay-W, Jeandré du Toit, Jedi94, Jeff G., Jeffrey O. Gustafson, JeffreyN, Jengod, Jennica, JeremyA, JesseRafe, Jethro Baggins, Jheiv, Jhsounds, Jhw57, Jim Huggins, Jim10701, Jim1138, JimVC3, JimboB, Jimknut, Jinnspirit, Jmg124, Jnestorius, JobSmee, Joeypk12, Jogers, John, John DiFool2, JohnOwens, Johnmoff82, Johnnya2k6, Jonathunder, Joniscool98, Jordan Elder, Joseph A. Spadaro, Joshua H-Star-R, JoshuaZ, Joymmart, Jraytram, Jskdpm, Jsmith12345, Juancameneses11, Julia14, Juliancolton Alternative, Jusjih, JustAGal, Justforasecond, Juzeris, Jwoodger, Jwrosenzweig, K ideas, KD-G722, Kafziel, Kaldari, Kanabekobaton, Katemorganishere, Kbdank71, Kchishol1970, Kelly Martin, Kennylucius, Kermitmorningstar, Kevin j, Kevindkeogh, Khenrich719, Kicking222, Kilo-Lima, KingoftheDogs, Kingpin13, Kinkyturnip, Kirill Lokshin, Kitch, Kollision, Kompik, Korg, Koyaanis Qatsi, Krich, Kripkenstein, Ksaraf, Kuddenchristo, Kumar Appaiah, KyraVixen, LHCpioneer, LOL, LaVidaLoca, Lady Aleena, Laggan Boy, Laser brain, LedgendGamer, Leewardwinds, Lena23supreme, LeoDV, Leszek Jańczuk, Lethe, LetsGoAngels, Leujohn, Levineps, Lhademmor, Liberlogos, Liface, Lightmouse, Limetolime, Lina Kd, Lionclaw, Liyah lover01, Logan, Loonymonkey, Lord Emsworth, Lord Eru, Lord Labak Daas, Loren36, Lovelovefilm, LovelyLillith, Lowellian, Luckyluke, Luckystars, Lukemogan, Lwalt, Lylenorton, MER-C, Maccoinnich, Madcapsjs, Madhero88, Madmardigan53, Magioladitis, Magmafiaman, Magnoliasouth, Majalinno, Majora4, Majorclanger, Malickfan86, Mandel, Mani1, Manuel Trujillo Berges, Manupa, Marek69, Mark91, MarnetteD, Martarius, Martin451, MartinDK, MartinRe, Martinwilke1980, Masaruemoto, Matt.whitby, Mav, Mazdapickup89, Mb1000, McSly, Meelar, Meeve, MegX, Menchi, Meno25, Mentifisto, Mepolypse, Mercury, Meredith Rodney McKay, Merqurial, Metafact, Mhrogers, Michael Farris, Michael J, MichaelCaricofe, Mikko Paananen, Mild Bill Hiccup, Minimac93, Minnesotajohn29, Miranda, MisfitToys, Miss Dark, Missy1234, MisterHand, Mito1970, Mm40, Moogwrench, MoraSique, Mormegil, Morwen, Mr Hall of England, Mr Stephen, MrDimples, MrDolomite, Mrceleb2007, Mrobamakid, Mrschimpf, Mrwill819, Msr69er, Mtmelendez, Mtrstore, Munin, Musicking, Mwalimu59, Myscrnnm, N5iln, NWill, NYScholar, Nahbios, NameIsRon, Nard the Bard, Natalie Erin, NawlinWiki, NeatlookingGirl, NellieBly, Neo The User, Neo-Jay, NeoAdonis, NeonNero, NickGarvey, Nightscream, Nihiltres, Nistra, Nixeagle, Nleamy, Nlu, Nmpenguin, Nologreenman, Noroom9, Noroton, Northamerica1000, Notheruser, Novangelis, NuShrike, Nuggetboy, Numbo3, Nutcracker, Nutshell8, Nzlmbrino, OLBKatz, ONEder Boy, Oanabay04, Obina, Oblivious, Ocean57, Off2riorob, Ohconfucius, Oknazevad, OldakQuill, Oleg Alexandrov, Olya, OneVeryBadMan, Onejaguar, Onorem, Oonml29, Orfilms, Orfy, OsamaK, Oscarsenvelope, Otolemur crassicaudatus, Ottawa4ever, Otto4711, Oughgh, OuttaHand22, Over24, Owen, Ozone009, Ozzieboy, PBP, PFHLai, PacificBoy, Pal400, Pamri, Passage987, Passargea, Pathivu, PatriceNeff, Patrick, Patrycja.pol, Patsycarpenter, Paul A, Paul August, Paulg99, Pegship, Pellicanshit, Pen1234567, People101, Peteferrara, Peter Isotalo, Peter S., Pethan, Phe, Phearson, PhilKnight, PhoenixMourning, Piano non troppo, Pieter Kuiper, Pigman, Pimlottc, Piragath, Pistachio disguisey, Pixi Uno, PlutoYellow, Pmjones 98, Pointer1, Pokemontrouble, Ponyo, Postdlf, Pperos, Ppk01, Ppntori, Princess Lirin, Proscribe, Puffy jacket, Puppy linuks, PurpleRain, Purslane, PuzzletChung, Pvhasure, Qaqaq, Quadell, Queenkissy, Quentin X, Quidam65, Quidom, Quinwound, QuiteUnusual, Qutezuce, Qwayfe, R'n'B, RBBrittain, RCMdel14, RS485, Rahrah89, Rajah, Rake, Ral315, Rami R, Rapigan, Ravothomas, Rdsmith4, Reaper Eternal, Reaverdrop, Reconsider the static, Red4tribe, Redspork02, Rekakaka, Removalteam, Resurrectionist II, Rettetast, Rex Momo, RexRey, Reywas92, Riana, Rich Farmbrough, Richhoncho, RickK, Riley Huntley, Rjanag, Rjwilmsi, Rl, Road Wizard, Robert Brockway, Robert Merkel, RobertG, Roberta F., Robma, Robsurfs, RockMFR, Rogerknauer, Romtobbi, Ronaldreagen80-88, Ronjohn, Rothorpe, Rowan Moore, RoyBatty42, Russeasby, Rusty, SEGATime2, SXD16, Saculаramacal13, Sam, Sam Li, Sango123, Santoshd, SarahLovesCats, Sarav online, Saw1998, SaxofoneDL, Saziel, Sceptre, Schmiteye, Schweiwikist, Scorpion0422, Scottfburg, Scrub121, ScudLee, Scwlong, Scyker, Seattlenow, Secret Saturdays, Seduisant, Seicer, Seraphimblade, Sf46, Shabidoo, Shadowjams, Shanel, Sherryon, Shirtwaist, Shniken, Shoefly, ShortShadow, Shreevatsa, Shymian, SilentMonkeys, Silentjames, Silverhorse, SimonP, Sin-man, Sintonak.X, Sjones23, Skier Dude, Skookie5, Slgrandson, Slikstar, Smartsitter, Smedpull, Smithereen, Sms231, Sn0wflake, SnakeChess5, SnappingTurtle, SnowFire, Snowman Guy, Snowolf, SoWhy, Soarhead77, Softlavender, Sole Soul, Somanytictoc, Son of Somebody, Sonam8311, Sonenreb1991, SoniC, Sonofike, Sonyack, Sooc2nd, Soumyasch, Spangineer, Spd4evr, Specialclifford, SpikeJones, Spoonkymonkey, SpuriousQ, Squandermania, Sqweak, Srich32977, Sry85, Stanley011, Starcheerspeaksnewslostwars, SteinbDJ, Stellmach, Steven Andrew Miller, Stickguy, Stlsportsfan2316, Stultae, Suffusion of Yellow, Suicidalsnowman, Summerleo, SuperOmegaStereo, Supermike6, Sus scrofa, Swaterfall, Swliv, Synergy, Szyslak, TARDIS, TAnthony, TGTIBTEE, Tadas12, Tailkinker, TallulahBelle, Tassedethe, Tbay01, Tbhotch, Technopat, Ted Wilkes, Teiladnam, Telesine, Tesseran, Th3perfectdrug, ThaCarter IV, The Wrong Man, The Zaniak, The-Pope, TheLastAmigo, TheLoverofLove, TheP3rfectDrug, TheRevolutionist, TheRingess, TheSBKdude, TheSuave, Theironduck, Thenature, Theoscarman, Thesilverbail, Theverysneakybat, Thief12, Thingg, Thobphon, Thomas Blomberg, Thores, Thorismund, Thumperward, Thuvan Dihn, Tide rolls, Tientao, Tiggerjay, Tillman, Tim Chambers, Tim1357, TimeLord, Timo Honkasalo, Tingrin87, Tinmanic, TinyMark, Tjmayerinsf, Tkgd2007, Tkreuz, Tkynerd, Tom.k, Tombobjones, Tomeasy, Tony Sidaway, Tony2Times, TonyTheTiger, TonyTonyTony, Towsonu2003, Trajce-Arsov, TransUtopian, TreyGeek, Tristan 753, Triwbe, Trusilver, Truthstudent, Tyler, UB65, Uber Cuber, Ugen64, Uncle Dick, Unclemikejb, UnfriendlyFire, Unni.0p, UnsourcedBlanker222, Unsy770, Uranium grenade, Ursinism, User27091, VMS Mosaic, Valley2city, VasilievVV, Veinor, Vgy7ujm, Vicki Virago, Vidshow, ViperSnake151, Vnowbee543, Vshahuk, Vsmith, W!ckedMe300, W!kiuser91, W.F.Galway, WOSlinker, Wack'd, Wahkeenah, Wain9309, Wakuran, Wallamaarif, Wangi, WeaselADAPT, Welsh, Werdna, WestJet, Westee, WhartoX, WhisperToMe, Wholphin, Widefox, Wiki alf, Wiki13, Wikievil666, Wikipedical, Wikipelli, Wikiuserphil, Wildhartlivie, WillMak050389, Willking1979, Wjsstray, Wknight94, Wlbatty, Wolcott, Wombatcat, Wonkeythemonkey, Woohookitty, Xanchester, Xoloz, Xymostech, Yaksar, Yamla, Yidisheryid, Yoddal, Yorkshiresky, Younggun1999, YoyaDiata, Yrgh, Zacdilone, Zanimum, ZenCopian, Zleitzen, Zoe, Zoicon5, Zoso Jade, Zosodada, Zotdragon, Zr123, Zundark, Zzuuzz, Zzyzx11, Δ, Александър, , , , 2264 anonymous edits

Fiorello_La_Guardia *Source*: http://en.wikipedia.org/w/index.php?title=Fiorello_La_Guardia *Contributors*: 4wajzkd02, AHMartin, AMittelman, AN(Ger), Abberley2, Abulanov, Akadruid, Alansohn, Alexander Tendler, Alexcoldcasefan, All Hallow's Wraith, Andyluciano, Anthony Appleyard, Anthony22, Appraiser, Arteitle, Arthena, Athene cheval, Aude, B.Wind, Balleur, Barticus88, Battlebeast, Bearian, BenedictX, Beyond My Ken, Bill Thayer, Binarybits, Biruitorul, BjornVDM, Blake-, Blanche of King's Lynn, Blue387, Brewcrewer, Briaboru, Brumon, Brutaldeluxe, CIreland, CWenger, CanadianLinuxUser, Captain Caveman, Charlene.fic, CharlotteWebb, Chorrocks, Choster, CliffC, Coldwarshot, Cometstyles, Commander Keane, Connormah, Conradege, Conversion script, Copyeditorjoe, Cr7i, Craverguy, Czrisher, D6, DBaba, Danny, Davepape, Deansjunk, Deflective, Delaywaves, Denisarona, DerBorg, Desempates, Dinopup, DocWatson42, Doctorfluffy, Docu, Donlarry, Doodle77, Dougie monty, Dr who1975, Dvd Avins, Edkochlives, Eino81, Ekotkie, Elmarco, Emilyvinson, EoGuy, Eranb, ErebusMyriad, Farras Octara, Floaterfluss, Foulfoul50, Froid, Gareth E Kegg, Genista, Gezellig, Gilliam, Giraffedata, Gobonobo, Goldenshepherd, Good Olfactory, Gosgood, Greekboy12345er6, Griot, Ground Zero, Halibutt, Happyme22, HarryCarney, Heddyhed, HennessyC, Hinotori, Hmains, HollyAm, Hughcharlesparker, Ichwan Palongengi, Ipeirotis, Iridescent, JForget, Jack O'Lantern, JackO'Lantern, Jake14001, JasonAQuest, Jerry Jones, Jhobson1, Jim.henderson, Jimp, Jmkleeberg, JoelShprentz, John Nevard, Johnpseudo, Joseph Solis in Australia, Joy, Joyson Konkani, Jtdirl, Jude23, JustAGal, K1Bond007, KF, Kaisershatner, Kazi2000, Kbdank71, Kchishol1970, Kingturtle, Kirby 456, Klemen Kocjancic, Koavf, Konczewski, Koplimek, Krao1119, Kross, Kumioko (renamed), KuroiShiroi, Kyle Barbour, L.eagleburger, LePoissonDeNoel, Leifern, Leonardo Alves, Lotje, Lst27, Luctor IV, Lukevl, MBisanz, Mark, Matthew Fennell, Mav, Mayumashu, Mervyn, Metal Militia, Michael Snow, Mike Garcia, Mike.daum, Minesweeper, Moncrief, Monkey Bounce, MrPrada, Mskyrider, Muboshgu, Mwanner, Mxn, Neutrality, Nivix, Nobunaga24, Nothing149, Nricardo, Nunh-huh, Oanabay04, Ohconfucius, Olivier, Onkel Dittmeyer, Pacholeknbnj, Padraic, PaulHanson, Pburka, Pfrishauf, Phyllis1753, Pl07442, Plaws, Polynova, Postdlf, Pvmoutside, QuantumOne, Qwerty0, RJFF, Rich Farmbrough, Richard Arthur Norton (1958-), Richard David Ramsey, Richard Weil, RickK, RiverStyx23, Rjanag, Rjensen, Rjwilmsi, Rockhopper10r, Rodney Boyd, Romanista, Rontrigger, Roswell native, SLMarcus, SPUI, Sadistik, Sander123, Sardanaphalus, SarekOfVulcan, Schmendrick, Sekicho, Selket, Sfaustin3, Sflorman, Shakescene, Shlomke, Sholom, SidP, Sk5893, Skywriter, Sloman, Smarticvs, Solace098, Someone else, Srich32977, Srnec, Sshikora, Station1, Stefanomione, SteinbDJ, SteveStrummer, Steveja, Sun Creator, Surfeit of palfreys, Tabletop, Tad Lincoln, TastyPoutine, Ted Wilkes, Teneriff, The Cipher, The truth about hephaestos, Thf1977, Thismightbezach, Tim bates, Timeineurope, Tiziano-Dossena, Tiziano52, Tom, TonyTheTiger, Tothebarricades.tk, Tragic Baboon, Tregoweth, Trojancowboy, Twintone, Tysto, UKoch, Ulric1313, ValenShephard, Vegaswikian, VerruckteDan, Vfp15, Viator slovenicus, Volvodd, Vulturell, Wechselstrom, WhisperToMe, Wikiklrsc, Wikiz876, William Allen Simpson, WojPob, World war two buff, Wormantson, XLR8TION, Xanzzibar, Xyzzyva, Yarnover, Zimbabweed, Zzyzx11, 'Ο οἶστρος, 350 anonymous edits

Uta_Hagen *Source*: http://en.wikipedia.org/w/index.php?title=Uta_Hagen *Contributors*: Abrazame, ActorScholar, Alex Middleton, AlistairMcMillan, All Hallow's Wraith, Angela, Aristeides, Asbestos, Ato, Avoided, Bardsandwarriors, Bbsrock, Belovedfreak, Biruitorul, Bovineboy2008, Bowalsh, Calton, Courcelles, Cuppysfriend, Cybninja, D6, Dante Alighieri, DionysosProteus, Download, Drpickem, Earthlyreason, Edgemar, Ellsworth, Emerson7, Ethylyoohoo, Ewlyahoocom, Ganymead, GenkiNeko, Grim Reaper (usurped), HOT L Baltimore, Here2fixCategorizations, Hu12, Icedude999, Ineuw, JEM1958, JGKlein, JJstroker, JamesAM, Jay-W, Jayepaul, Jeanenawhitney, Jfreer, Jivecat, Jmkleeberg, Johnpacklambert, JonGwynne, Joseph A. Spadaro, JustDerek, Karas peter@yahoo.com, Katharine Reszetnik, Killiondude, Koavf, Kyle Barbour, Lars T., Lilac Soul, LiniShu, Marcuse, Mattbr, Mbdfs, Michael David, Moeron, Natalietrenlay, Noirish, Olessi, Owen, PDH, Paul A, Pissant, RFD, Rentaferret, Rentstrike, Rjwilmsi, Rms125a@hotmail.com, Rossrs, Rrburke, SDC, Sam, Sciurinæ, Sgactorny, ShelfSkewed, Steve Casburn, Steven J. Anderson, Superbob10, Ted Wilkes, The Cunctator, The wub, Thumper08, Tinton5, TommyP, Tree Trimer, Vulturell, Walloon, Walter Breitzke, Wik, Will Beback, Willking1979, Wulf Isebrand, Xezbeth, 142 anonymous edits

***Winged Victory* (play)** *Source*: http://en.wikipedia.org/w/index.php?title=Winged_Victory_%28play%29 *Contributors*: Ali'i, Aristophanes68, Cbl62, Cla68, Fuhghettaboutit, Jbmas99, Jezhotwells, Kateshortforbob, Kauffner, Monkeyzpop, Mr Hall of England, Tassedethe, Tenor2bass, Woohookitty, 6 anonymous edits

Mario_Lanza *Source*: http://en.wikipedia.org/w/index.php?title=Mario_Lanza *Contributors*: 4meter4, AKeen, Aardvarkzz, Academic Challenger, AlbertSM, Alfietucker, Algont, All Hallow's Wraith, Alphageekpa, Amsterdave, Andonic, Antandrus, BaronLarf, Belovedfreak, Belvederehedon, Bgwhite, Bjankuloski06en, Blue Tie, Bluetooth954, Bratsche, Brian Joseph Morgan, Bundas, CIreland, Callmarcus, Camillus McElhinney, CanisRufus, Cathytreks, Cernen, Chanlyn, Clarityfiend, Clausule, ClockworkLunch, Cmprince, Condoleeza, Copana2002, Cresix, Crystallina, Curtaintoad, D6, Danny, Derek R Bullamore, Derekmcgovern, Design, Discospinster, Dissolve, Doniago, Dutchy85, Dweaver52, Edcolins, Elmarco, Ergonaut2001, FrankEldonDixon, Gcstackmoney, Geodyde, Ghirlandajo, GiantSnowman, Grm wnr, Guv2006, HOT L Baltimore, Hex valant, Hmains, Hu12, IGG8998, Infrogmation, Ionutzmovie, Irishguy, Jack O'Lantern, Jack Oldcastle, JackofOz, Jag123, Jerrybarker, Jim Simpson, Jimmtpicornell, JorisvS, Joseph A. Spadaro, Joyradost, Junglejill, Kakofonous, Kaneshirojj, Karas peter@yahoo.com, Kbdank71, Kedar63, KeithH, Kenatipo, Kkmurray, Kleinzach, Ksnow, LaDanza, Labalius, Lashuto, LilHelpa, Lpgeffen, Lugnuts, Lynnwall, Makemi, MarnetteD, Mattbr, MauriceReeves, Mike 7, Mike Dillon, Milstein2001, Miraclewhip, Monkeyzpop, Moreschi, Natashacatrina, NewYork1956, NiceguyC, Nishkid64, OLEF641, ObjectiveBrit, Okc, Onefortyone, Orange Suede Sofa, PMDESQ, Paul A, Paul MacDermott, Pinkadelica, Rigadoun, Rjwilmsi, Rms125a@hotmail.com, Rocks and rolls, Rossrs, Rpstuvabc, Rrose Selavy, Ruedetocqueville, Sam L. Camp, ShelfSkewed, Sillyfolkboy, Solitude, Spidermine, Srich32977, Stan Shebs, StealthStar, Stefanomione, Sus scrofa, Tacitus Prime, Tamfang, Tanet, Tassedethe, The Thing That Should Not Be, Threeafterthree, Timrollpickering, Tjmayerinsf, Tnxman307, Toccata quarta, Tony1, Tonytenor, Ugen64, Vedran12, Viajero, Violncello, Voceditenore, W guice, WWGB, Wallie, Wikiwatcher1, Wildhartlivie, William Avery, Woohookitty, Wormygermy, Xenon54, 285 anonymous edits

John_Forsythe *Source*: http://en.wikipedia.org/w/index.php?title=John_Forsythe *Contributors*: Aardvarkzz, Abrazame, Alansohn, All Hallow's Wraith, AllyUnion, Allycat, Anthony Winward, ApolloCreed, Ary29, BSTemple, BajanZindy, Ben King, Bender235, Billy Hathorn, Biographies 1, Bobo192, Boleyn2, Bond007rh, Bridgetfox, Budxvg1, CARTMANSTEWIESCOTT, CRJ200flyer, Cadsuane Melaidhrin, Canadian Paul, CanisRufus, Cavarrone, Clarityfiend, Closedmouth, Cocoruff, Colonies Chris, D6, DandyDan2007, David L Rattigan, DavidESpeed, Dbenbenn, DemirBajraktarevic, Demiurge, Dfmock, Djinn112, Dmlandfair, Dnwyfc, Eclipse13, Efrem7, Ejfetters, Fabiform, Favonian, Feezo, Finalius, Firsfron, Fordmadoxfraud, Forsyth96, Gaius Cornelius, Gcollinsii, Gcollinsii@aol.com, Gcstackmoney, Gibb0Yeo, Gilliam, Gran2, Green Gables, Guerillero, Gwen Gale, HDS, Handicapper, It7782887, Italianlover07, Ixfd64, JGKlein, JackofOz, Jamesbanesmith, Javierito92, Jcbarr, Jcsrauch, Jdebenedict, Jeanenawhitney, Jeff3000, Jeffq, Jfg284, Jim Michael, Jimfbleak, Jimknut, Jkaharper, Jogers, John Anderson, Joshdboz, Kanabekobaton, Katydidit, Keane4, Kilowattradio, Koavf, Kokomo Coconuts, Koplimek, Kostelca, Kumioko (renamed), Lajsikonik, Logical Fuzz, Lord antares, Lugnuts, Marcika, Mark Sublette, Masterknighted, Mattbr, Mel Etitis, Michael Hardy, Mike Halterman, Mild Bill Hiccup, Monegasque, Monkeyzpop, Moviefan, Mushroom, Music2611, Neelix, Nghtownclerk, Nixeagle, Noirish, Obituarist, Peelsyadot, Philip Cross, Pinkadelica, Pizzamaniac09, Ponyo, Postcard Cathy, R'n'B, Rms125a@hotmail.com, Roaming27, Rossrs, Sam, Sandra.schellhase, Savolya, Secret (renamed), ShelfSkewed, Shell Kinney, SigKauffman, Sky83, Srikeit, Steam5, Sukee, TAnthony, TFBCT1, TMC1982, Tassedethe, The Ink Daddy!, TheCustomOfLife, TheRhani, Thismightbezach, ThomasHarte, Tinton5, Tjmayerinsf, Trut-h-urts man, Ulric1313, Vincelord, Vsmith, Way2rude, Waylonrobert, White 720, Whywhenwhohow, Wildhartlivie, Wilfried Derksen, Wizardman, Wtimrock, YUL89YYZ, Zanaq, ZouBEini, Zzyzx11, 196 anonymous edits

Karl_Malden *Source*: http://en.wikipedia.org/w/index.php?title=Karl_Malden *Contributors*: .anaconda, 23skidoo, Aaron Bowen, Abrazame, Absalom89, Aca Srbin, AdamBMorgan, AdjustablePliers, Agamemnon2, Aldis90, All Hallow's Wraith, Ameliorate!, Anthony Winward, Archiefhist, ArgentumOfOz, Arvidsponberg, Asalrifai, Astorknlam, Astrostart, BRG, BSTemple, Bbik, Belovedfreak, Bg007, Bgwhite, BigGuy, Bigobigo, Billyfutile, Biruitorul, Bjones, Bongwarrior, Booshakla, Bryates999, Btphelps, C.Fred, Canadian Paul, CanisRufus, Captain Kirk, Careful Cowboy, Carski, Cburnett, Ceeforcat, Cerejota, Closeapple, Cnbajr, Computar, Correogsk, Cosprings, CraigRNielsen, D6, DBaba, DGFS6691, DGaw, DJ Bungi, DStoykov, Dado, DanielRigal, Darwinek, DaveGorman, David Gerard, DavidESpeed, Dcabrilo, Deb, Deliboo24, Deville, Dirtpig, Djordjes, Dml, DocOctopus, DocWatson42, Dreyfus, Duncanbruce, Echtner, Edcolins, Edward321, Ekwtsm, Emerson7, Esrever, Evlekis, Ewlyahoocom, Fadookie, Failte, Fdewaele, Fernandosmission, Frank, Frecklefoot, Fryede, Garion96, Gcollinsii, Gcollinsii@aol.com, Gcstackmoney, Gmchambless, Good Olfactory, Gorrillla5, Gr8lyknow, Gradanin, Granpuff, Great Scott, Griffinofwales, GusF, Gustav von Humpelschmumpel, HADRIANVS, Hadžija, Happyme22, Hmains, Holdenbuckley, HollyAm, Hydrargyrum, Igor, Information yes, Infrogmation, Irregulargalaxies, Italianlover07, J.delanoy, JGKlein, Jacek Kendysz, Jack Cox, Jack O'Lantern, JackofOz, JamesMadison, Jan Arkesteijn, Japanese Searobin, Jeanenawhitney, Jeff3000, Jeffconn, Jeffrey Smith, Jellybaby 344, Jetman, Jguad1, Jim Michael, Jjj055, Jkaharper, Joegoodfriend, Joey80, John, John lippiatt, John of Reading, Johnpacklambert, Joshmaul, Juju 2402, Jusdafax, Justme89, Kaaveh Ahangar, Kauffner, Kbdank71, Kintetsubuffalo, Koavf, Kumioko (renamed), Kusunose, Lajsikonik, Lambejim, Lauracameron, LeeNapier, Leon7, LittleOldMe, Lockley, Lothar97, Lugnuts, MK013, Man vyi, MarnetteD, Martarius, Masterknighted, Mayumashu, MegX, Meribona, Michael Hardy, Mike Dillon, Mile32, Mladen George Sekulovich, Monkeyzpop, MorganaFiolett, NOLA504ever, Nabokov, Nasnema, NawlinWiki, Necrothesp, Nemonoman, Nickellmusic, NightMonkey, Nikola Smolenski, Ninam, Nlu, Noirish, Noviracer, Ntsimp, Oatmeal batman, Obituarist, Oobopshark, Original Oracle, Orokrstas, PANONIAN, PCR2000, PaulinSaudi, Phbasketball6, Philip Cross, Pizzamaniac09, Postcard Cathy, Prodego, ProhibitOnions, Pwt898, Qertis, Queenmomcat, Quentin X, Rade Nagraisalovic, Rjwilmsi, Rmosler2100, Robneyer, Rogerd, Rossrs, Runcorn, S, Saforrest, SarekOfVulcan, Scanlan, Scarce, Scriberius, Scwlong, Sekula777, Shadow2700, ShelfSkewed, Shortie43504, Sift&Winnow, Simetrical, Skol fir, SkymasterUK, Slulek, Snake bgd, Spacini, Spanglej, Stainlessboy, SuperMarioMan, Surazeus, Sverdrup, TDKR Chicago 101, Tassedethe, Tbhotch, Tcmcelmurry, Ted Wilkes, Tenar80, TexasAndroid, The Man13, The Thing That Should Not Be, TheGeekReport, Thismightbezach, Timmyo101, Tinton5, Tiusp, Tjmayerinsf, Tkreuz, Tom, TonyTheTiger, Treybien, Tyger63, V.j.sekulovic, Vanjagenije, Vegaswikian, Vicenarian, Vitriden, WTucker, WWGB, Warrior4321, Wax, Wernher, Where, Wildhartlivie, Willerror, Willking1979, Wmahan, Woohookitty, Yakushima, Yekrats, Yokes14, Zarateman, Zeus1234, Zoicon5, Zureks, Zzyzx11, 387 anonymous edits

Lee_J._Cobb *Source*: http://en.wikipedia.org/w/index.php?title=Lee_J._Cobb *Contributors*: 2002:1814:17F3:1234:226:8FF:FEE2:BADD, Aardvarkzz, AlanM1, AlbertSM, Alcmaeonid, Aleal, All Hallow's Wraith, Apikoros88, Arclen, Ary29, Attilios, Billy Hathorn, BlackJack, BrianBoru10, Burgundavia, CXCV, Canis Lupus, Cavarrone, Cburnett, Clarityfiend, Coffee, D6, Dbart, DemirBajraktarevic, DisillusionedBitterAndKnackered, Donmike10, Dr. Blofeld, ERJANIK, Erik9, Etaicq, Firsfron, Formeruser-81, Fru1tbat, Garion96, GregorB, Grenavitar, Gustav von Humpelschmumpel, Hmains, Hokeman, Hydrargyrum, IZAK, Ikre8, Inwind, Iowaway, J. Van Meter, JJstroker, JJstrokey, Jack Merridew, Jack O'Lantern, JackofOz, Jason Palpatine, Jbmas99, Jeanenawhitney, Jeffman52001, Johnpacklambert, Jonathan.s.kt, Jordanceltics fan, Judgeclay, Kate, Kukini, Kumioko (renamed), Lightmouse, Lionelt, Lockley, Lord Cornwallis, Lugnuts, MarnetteD, Martinl, MegX, Megan1967, Michael David, Missy1234, Monkeyzpop, MovieMadness, Netscott, Niteowlneils, Noirish, PKT, Pegship, Phbasketball6, Pinkadelica, Postcard Cathy, RedfOx, Riteshkapoor, Rossrs, SFTVLGUY2, Salamurai, Sam, Savolya, Sbeath, Scanlan, Stetsonharry, Storyliner, Sumahoy, Tallyho70, Tangurena, Tassedethe, Ted Wilkes, Thismightbezach, Tjmayerinsf, Tomticker5, TonyW, Tsujigiri, Ulric1313, Volatile, W guice, Wars, WebHorizon, WikiPuppies, Wildhartlivie, Winstonwolf33, Woohookitty, Zone46, 70 anonymous edits

Darryl_F._Zanuck *Source*: http://en.wikipedia.org/w/index.php?title=Darryl_F._Zanuck *Contributors*: Aitch Eye, All Hallow's Wraith, Aloysius the Gaul, Anonymous from the 21st century, Antandrus, Anthony Winward, Aymankamelwiki, Azucar, Barkeep49, Belmontian, Bill Thayer, Bzuk, Carelesskiss, Cholling, Commander Keane, ConradKilroy, Darwinek, DemirBajraktarevic, Dthomsen8, DuncanHill, Ed Poor, Elchinsu, Empoor, Everyking, F W Nietzsche, Foofbun, Freako, FredR, Gettingtoit, Ghosts&empties, Gingerandfred, Granpuff, Gyonis, Hephaestos, I'mMe!!,

IP4240207xx, Iwillbeamoviestar, JGG59, JGKlein, JaGa, Jack O'Lantern, JillandJack, Jls068, Johnpacklambert, Jonathan Stokes, Josh Rumage, Jusdafax, Kasper2006, Killuminator, Klantry01, Koplimek, Kumioko (renamed), Lambiam, LorenzoB, Louklou, Lugnuts, Maltmomma, Martarius, Matthew Fennell, Meribona, Michaelt54, MisfitToys, MisterMorton, Mr Hall of England, Nasnema, Noirish, Novasource, PedanticallySpeaking, Peter Greenwell, PhantomS, Philip Cross, QYV, Rebel Yeh, Richard Arthur Norton (1958-), Rms125a@hotmail.com, Rossrs, Ryanasaurus0077, Savolya, Selbymayfair, Sfahey, Sideshow Bob Roberts, Spanglej, Spot87, Stefanomione, TDKR Chicago 101, Ted Wilkes, The Nuke, Thepinterpause, Thumperward, Tjmayerinsf, TonyTheTiger, TwoOneTwo, Vgirly, Viriditas, Vulturell, Wheelie Willy, Wulf Isebrand, Yinzland, Zanimum, 81 anonymous edits

Winged Victory (film) *Source*: http://en.wikipedia.org/w/index.php?title=Winged_Victory_%28film%29 *Contributors*: Ary29, Fpastor, Jbmas99, Monkeyzpop, Pastor Theo, Sus scrofa, Varlaam, 1 anonymous edits

George_Cukor *Source*: http://en.wikipedia.org/w/index.php?title=George_Cukor *Contributors*: 15.253, 20yearoldboyfromNY, Aardvarkzz, Adam78, After Midnight, Ahoerstemeier, Akcarver, Alansohn, All Hallow's Wraith, Anthony Winward, Arch dude, Asparagus, Astorknlam, Aztek888, Azucar, BD2412, Babək Akifoğlu, Bbsrock, Bearcat, Bellhalla, Betempte, Blackjanedavey, Blahomfg, Bzuk, Chandler75, Chotterap, Classicfilmbuff, Cmulrooney, Coder Dan, ConcernedPhotographer, ContiAWB, Conversion script, Crumbsucker, Cskor, Cst17, D6, Davshul, Db4260, Dcs315, DeansFA, Deb, Deflective, Dr. Blofeld, DrBat, Ellsworth, Emerson7, Enchanter, Estrose, ExpatSalopian, FisherQueen, Fratrep, Frenchjr25, Fuhghettaboutit, GJPelisson, Gareth E Kegg, Gcstackmoney, Giebergoldfarb, Gkklein, Gogo Dodo, Gwendal, Gwern, HennessyC, Highlunder, Hmains, Hobartimus, Ianblair23, Ilion2, Infrogmation, JGKlein, JYing, Jack Merridew, Jack O'Lantern, JackO'Lantern, JackofOz, Jbmas99, Jburlinson, Jeremy Butler, Jeronimo, Jfiling, JillandJack, Jim.Mignano10, Joao Xavier, Joey80, John Reaves, John Smythe, Juko, Kbdank71, Kbthompson, Keane4, Kittybrewster, Koavf, Koyaanis Qatsi, Kumioko (renamed), Kwamikagami, Lilac Soul, LiteraryMaven, Look2See1, Lugnuts, Lynchkenney, MWaller, Magioladitis, Manuel Trujillo Berges, Marb, Martarius, Matty j, Mav, Mayumashu, MegX, Midnight man, Mike Dillon, MisfitToys, Monkeyzpop, Morn, Mymansyd, Nick, Nick Number, Noirish, NotACow, Notepadzone, Orbicle, Otto4711, Ozgod, P. Rollin, PacificBoy, Passionvine, PhantomS, Philip Cross, Pibwl, Piccadilly, Porlob, Proscript, QuickMotion, RBBrittain, RafikiSykes, Ravenclaw, Rcs1898, Remember, Rodasmith, Rossrs, Ryulong, S2grand, Sam, Ser Amantio di Nicolao, Sgeureka, ShelfSkewed, Shortride, Sophitus, Stolengood, SubSeven, Svsvtkag, TFBCT1, Tassedethe, Ted Wilkes, TheMadBaron, Thefourdotelipsis, Thepangelinanpost, ThylekShran, Tinton5, Tjmayerinsf, Treybien, Twix1875, TwoOneTwo, UZiBLASTER7, Ulric1313, Uparedes, Vaganyik, Vobor, Vulturell, Walloon, Wavelength, Wildhartlivie, Woohookitty, X96lee15, Xanzzibar, YUL89YYZ, Yinzland, 103 anonymous edits

European_Theatre_of_World_War_II *Source*: http://en.wikipedia.org/w/index.php?title=European_Theatre_of_World_War_II *Contributors*: 30 SW, 42omle, Adel.M.Radwan, After Midnight, Ahoerstemeier, Altenmann, Ancheta Wis, Anotherclown, Art LaPella, Axlrosen, Bad Night, Barnhim12, Bassistphysicist, Bearcat, Binary TSO, Bjmullan, Black Falcon, Blacksheep2c, Bobblewik, Bobby D. Bryant, Brian Boru is awesome, Bruce Hall, CGreen, CORNELIUSSEON, Calliopejen1, Caltas, Camino, Can't sleep, clown will eat me, Capt Jim, Capt. James T. Kirk, CaptainFugu, Century0, Choihei, Chris 73, Chris the speller, ChrisO, Christofurio, Colonies Chris, CommonsDelinker, Constanz, Courcelles, Cpenikis, Crvena zvezda, Curtis Wong, Cyfal, D, DJ Clayworth, DO'Neil, Darry2385, David Kernow, David R. Ingham, Dcoetzee, Defender of torch, Delanie22374, Deltabeignet, Depp Jones, DexDor, Dhimwit, Dismas, Dna-webmaster, DocWatson42, Dodo19, Dpm64, Dwo, ESkog, EatAlbertaBeef, EddieVanZant, Edward130603, Edward321, Elk Salmon, Ellsworth, EnigmaMcmxc, Eric-Wester, Escape Orbit, Esn, Esprit15d, Evropaios, Excirial, Fabartus, Fairlightseven, Finavon, Fredrik, Frietjes, Funbeginshere12345, GCarty, Gareth Wyn, Garion96, GeneralPatton, Gob Lofa, GraemeLeggett, Grant65, Gugilymugily, Gwabell, Halibutt, Hmains, Ho2, Hohum, Husnock, Hutcher, Ibbn, IceKarma, IceUnshattered, Ioakinf, IstvanWolf, Itai, Ixfd64, J.delanoy, J04n, J8tennant, JLaTondre, Jaan, JamesAM, Jason M, Jimjoe, Jj137, Jmlk17, Jons63, Jooler, Joy, Jrleighton, Kelisi, KezianAvenger, Kiefer.Wolfowitz, Kirrages, Ko Soi IX, Korg, Kronnang Dunn, Kubanczyk, KuriosD, L Kensington, Lapsed Pacifist, Legolas3, Leoni2, Leszek Jańczuk, Levineps, LibLord, Light current, Lightmouse, Ligulem, Lilac Soul, Lisamt, Lohengrin1991, Lonelydarksky, MECU, MER-C, Mangostar, Manxruler, Mapakyinjamie, Marek69, Mark.murphy, Maseff2, Matt Crypto, Mattworld, Maxim, Mellum, Mentifisto, Mgedmin, Michael Hardy, Mike Young, Mikeo, Minesweeper, Mintguy, Miranche, Mkpumphrey, Mocninja, Morstar, Morwen, Mrg3105, Mugs2109, Mzajac, N328KF, Naddy, Naryathegreat, Neutrality, Nightenbelle, Nilmerg, Nmpenguin, Nneonneo, NorthernKnightNo1, Nyenyec, Oberiko, Oblivious, Obradovic Goran, Optimist on the run, Oxymoron83, PBS, PRhyu, Parthsaraiya, Paul Benjamin Austin, Paul Siebert, Peregrine981, Periklis*, Pigsonthewing, Piotrus, Plasma Twa 2, Proofreader77, Publicus, Qertis, QuirkyQuark, QuiteUnusual, Rainsford13, Rama, Random contributor, Raul654, RedWolf, Renski, Rich Farmbrough, Richard Keatinge, Rjensen, Rjwilmsi, Roary456, RobertG, Robertvan1, Roda, Romanm, Rotard4989, Rubisco, Ruhrjung, Sam Hocevar, Sam Korn, Seb az86556, SecretAgentMan00, Shadow007, Silverhorse, Simeon24601, Simishag, Spikebrennan, SpookyMulder, Ssolbergj, Starbuck2015, StillTrill, Styrofoam1994, Sylvar, THB, Tewirp, The Garden Gnome, The High Fin Sperm Whale, The PIPE, The Thing That Should Not Be, Thebeginning, Theopolisme, Timwi, Tobyw87, Tom, Tom-, Tony1, Toya, Tpbradbury, Tuomas, Usergreatpower, Vanished user 90345uifj983j4toi234k, Vanished user ewfisn2348tui2f8n2fio2utjfeoi210r39jf, Voidvector, Wallie, Wempain, Whiskey, WhisperToMe, White Shadows, Wik, Wiki0709, Wikid77, Wikipelli, Woohookitty, YaltaExpert1945, Yoninah, Zickzack, Zoicon5, 360 anonymous edits

Mickey_Rooney *Source*: http://en.wikipedia.org/w/index.php?title=Mickey_Rooney *Contributors*: 23skidoo, 2tuntony, 6afraidof7, 850 C, AVC User, Adam Carr, After Midnight, Al.locke, Alamandrax, Alansohn, AlbertSM, All Hallow's Wraith, Allehboo, Amarkov, Amaury, Amchow78, American Eagle, Americanhistorybuff101, Amxitsa, Anastrophe, And1987, Andyh2, Anonymous Cow, Anthony Winward, Anthony22, Arjoccolenty, Armymp0165, Aspects, Astynax, Avalock, Avono, Aytrus, B, BHillbillies, BRG, BVA, Barliner, BarryTheUnicorn, Beao, Bearcat, Before My Ken, Benjiboi, BillFlis, Billy Hathorn, Blair P. Houghton, Blue954, Bobblewik, Bondegezou, Bongwarrior, Bovineboy2008, Bovlb, BozzieBear, Bped1985, Bradleycanfield, Brenont, Bridgetfox, Bunnyhophop123, CWii, CallMeHenry, Canadian Skeptic, Captin Shmit, Casper2k3, Caulde, Cavarrone, Cburnett, Ccady, Chillum, Choster, Chris fardon, Chris.P.Dunning, Christophski1, Cilstr, Cinatyte, Clarityfiend, Cntras, Colonies Chris, Comedy3, Coyets, Crakkerjakk, Creepy Crawler, Cst17, Cunningpal, Curtsurly, D6, DARTH SIDIOUS 2, DBigXray, Dalexwats, Daljp11, Darwinek, David Rush, DavidFuzznut, DavidOaks, Dcsohl, DemirBajraktarevic, Dicklyon, Digby scallops, Diligent007, Discospinster, Dlambda, Doc Velocity, DoggyFizzle, Don1962, Donmike10, Downwards, Drew R. Smith, Droll, Dschroder, Dthomason07, Ed Poor, Edokter, Edony, Edward 75, Edward Wakelin, Edwy, Ejfetters, Emerson7, Enosfam, Epbr123, Erechtheus, Essexmutant, EvelinaB, Exeterjay, Facts707, Falcon Kirtaran, Fastily, Favonian, Fawcett5, Fendue, FlyingPenguins, Foofbun, Fourthords, Fram, Frank21212, Frankenpuppy, Frog47, Fuhghettaboutit, Fuzheado, GStreetwalker, Gamaliel, Ganymead, Gemini86 618, Geometricks, GetSmartEditing, Ghaleonh41, Ghirlandajo, Gibbynation, Glane23, Go Sox, Gobonobo, GodfreyDaniels, GraemeL, Gran2, GrayFullbuster, GreenGlass1972, Greswik, GusF, H Bruthzoo, HHyttinen, Hardy1956, Hedpeguyuk, Heritage727, Hifihitman, Hobbes Goodyear, Hollywoodfact, Holothurion, Home Row Keysplurge, Huangdi, Humdingers, Hypnoxscrum, IanManka, IceCreamAntisocial, Icez, ImbytheAsianSanta, Imnotminkus, Informationfountain, Ionesco, Iridescent, Italianlover07, J.Nevels, J.delanoy, JGKlein, JHP, JSweit8573, JaGa, Jack Cox, Jack O'Lantern, Jack rosell, JackofOz, Jaldridge86, James possamai, Jason Dickert, Jburlinson, Jeff3000, Jeffman52001, Jim1138, Jimknut, Joefromrandb, Jogers, John Nevard, Johnnybriggs, Jojhutton, Jonathanwallace, Josh Rumage, Jpgordon, Juowikis, JustAGal, Justinboyer, Karen Johnson, Karinrg, Kbdank71, Kchishol1970, Keane4, Kingstowngalway, Kingturtle, Kiril Simeonovski, Kumioko, Kumioko (renamed), Kurtfetherolf, Kwekubo, LBM, Landroo, Largoplazo, LarryHoward, LedgerJoker, Leileilol, Levineps, Lilredrooster, LittleHow, Lockley, Longhair, LtPowers, Lugnuts, Luna Santin, Lusiphur, Lyverbe, ManOfSummer, Mandsford, Manuel Trujillo Berges, Mark Sublette, Markpape0220, Marktreut, Marproductions, Martin8721, Marysunshine, Mastermike890, Materialscientist, Matt Deres, Matt-rex, Mattbr, Mav, MegX, Melos Antropon, Merqurial, Michaelsbll, Mickeyrooneyfan, Migp, Mike 7, Millahnna, Minor4th, Mmusselman, MovieBuff74, Mr Hall of England, Mr.Atoz, Mrsunen, Musdan77, Mutt Lunker, Myleslong, Mythic Writerlord, NHRHS2010, NTDOY Fanboy, Neddyseagoon, Nehrams2020, Neilc, Nem1yan, Nemeses9, NerdyScienceDude, Neumannk, Neverland1, Nghtownclerk, Nicholas0, NickCT, NicknameBoy, Ninjawarriordex, Noirish, Nosebutton, NovakFreek, Number2211, Number87, Nymf, Oanabay04, Olivier, Ollie354254, PKT, Patkirkwoood, Paul A, Paxsimius, Pearle, Pennyjack2, Personpersonperson, PhantomReinhart, Philip Cross, Philip Trueman, PhilipR, Phoenix Flower, Piggyjoe1234, Pilotguy, Pinethicket, Pinkadelica, Pizzamaniac09, Pointswestfan, Ponyo, Preschooler.at.heart, Pressedfortime, Princess Lirin, Professor Von Pie, ProveIt, Puppetguy, Qdiderot, QuasyBoy, Quebec99, R.h.c.afounder1, RJaguar3, RMc, RWD 2, Rancorbeast, RandomOrca2, Ravy, Rcsprinter123, Red blaze, Redeagle688, Renegade Replicant, Rich257, Richardsjp2, RjCan, Rms125a@hotmail.com, Robert Fraser, RobyWayne, Rockstarstylist, Rodhullandemu, Rogerd, Ronin Aquila, Ronweasley601, Rooney2615, Rooneyincomedian, Rossrs, Rrose Selavy, SQGibbon, Sage Sophia, Saworkman, Sceptre, Schmoore, Scoty6776, Seedbuddy, Selma Simpson, Shadow2700, Shannernanner, Shijjiri, Shirulashem, SimonATL, Sinclairindex, Skymasterson, Slgrandson, Slingbat, Softlavender, Solarfilmworks, Sp3lly, Spanglej, Sparky the Seventh Chaos, Spudroe, Srich32977, Ssilvers, Staffwaterboy, Status, Streltzer, Studerby, StudierMalMarburg, Swanny18, T. Anthony, TCY, TPIRFanSteve, Tabledhote, Tacothecat, Talory, Tassedethe, Teamplaya, The Giant Puffin, The wub, TheMitch1966, Thekingofcarryon, Think outside the box, Tim Thomason, TimGallagherActor, Timmyo101, Tinton5, Tjmayerinsf, Tpbradbury, Trackinfo, Trainfan01, Tregoweth, Trevor MacInnis, Trusilver, Tunes, UltimatePyro, Unother, Useight, Ustye, Valusa, Vedek Dukat, Versus22, Vitriden, Voxparadox, Vulturell, WTucker, We hope, Wedg, Whigham1, Why Not A Duck, Wikieditor06, Wikievil666, Wikipelli, Wikiwatcher1, Wildhartlivie, Will Beback, Wizzard2k, Woohookitty, Wrightjh, XRiamux, Xfpisher, Ylee, Zaphod-Swe, Zetterberg40, Zoggykrieger, Zscout370, 785 anonymous edits

Big_band *Source*: http://en.wikipedia.org/w/index.php?title=Big_band *Contributors*: 33mg1, ACupOfCoffee, Abronkeeler, Accordion Noir, Addihockey10, Ae2161, After Midnight, Aitias, Alansohn, Alarob, Aldis90, Alex.muller, Alexcat, AliaGemma, AllTalking, Allenstone, AllyD, AndrewWTaylor, Antandrus, Ariedartin, Atpal, Badagnani, Ballchin101, Bbq332, Ben Ben, Ben1220, Binksternet, Biopresto, Biosketch, Bobo192, Bonadea, BostonMA, Braindrain0000, Brequinda, Bydesignonly, C777, Camembert, Canterbury Tail, CattleGirl, Chevellefan11, Chinasaur, Circumgoy, Clarince63, CommonsDelinker, De mortuis..., Deathawk, Dissolve, DocWatson42, Dogru144, Dragonire, Dtmwtp, Eandboo2, Edison, Ejazzlines, Emersoni, Erikpoort, Ernie1, Evaa, EvilPizza, FOXTROT1924WB, Fawcett5, Fred the Oyster, Freechild, Gab.popp, George Leung, GregorB, Gurch, Guy Hatton, Gyrofrog, Hmains, Hornandsoccer, Husond, Hyacinth, Hyju, IKF, ItCanHappen, ItsAlwaysLupus, J.Rohrer, J04n, Jackknight28, Jamoche, Jansiana, Jason Quinn, Jeff3000, Jengod, Joel7687, Joke-wik, JonMoore, Jongleur100, Julien Tuerlinckx, KF, KINGLVTX300, Ka Faraq Gatri, Kabuto 7, Kintetsubuffalo, Kirachinmoku, Kisch, Kku, Klooka, Kyng, Leahtwosaints, Lewisjamesz, Light current, Lightmouse, LittleWink, Lozeldafan, LtPowers, Luca Pacioli, Luna Santin, Mac Dreamstate, Maczad, Maestrovox, Magioladitis, Mako098765, Markkawika, Masterpjz9, Mattingly23, Mboverload, MedicD, Megaman en m, Mel Etitis, Mephistophelian, Mercury McKinnon, Merphant, MiLBfan, Michael Devore, MichaelBillington, Mr.Xp, Mrtea, Myofilus, Mütze, Necrojesta, Nick Number, Nigelj, Nuance 4, Nv8200p, Oakie102, Oemmett, Omnipedian, OnBeyondZebrax, Oneiros, Pascal.Tesson, Paul August, Pensativa, Pepso2, Peter cowley, PhantomS, Pharaoh of the Wizards, Piano non troppo, Pjs012915, PrestonH, ProveIt, Pugetbill, Qxz, RadioBroadcast, RandomP, RedWolf, RekishiEJ, Rev Edward Brain, D.D., Rholton, Rich Farmbrough, Richhoncho, Rory77, RoyBoy, SD5, Sabrebd, Saltlakejohn, Sam Hocevar, Sardanaphalus, Saros136, Sarsdran, Scotwriter, Sirimiri, Sjhnumber1, Sluzzelin, Slysplace, Soulpatch, SpuriousQ, Stan Shebs, Stephensuleeman, Steve Jackson, Steve Pastor, StorkenSture, Student7, Studerby, Stutty, TUF-KAT, Tainter, Tarret, Tassedethe, The Hybrid, The wub, Thomsen, TimNelson, Tjmayerinsf, Tommstein, ToroQ, Troyastorino, Uncle Dick, Webchem, Welsh, Wine Guy, Witchwooder, Wmkvfm, Zazaban, 476 anonymous edits

Sayonara *Source*: http://en.wikipedia.org/w/index.php?title=Sayonara *Contributors*: AdamW, Adraeus, Aladin34, BiT, Bobet, Bobo192, Boneyard90, Brunhildr, Bryan Derksen, Calmer Waters, Cburnett, Cliff1911, CrypticBacon, Cybercobra, DaveGorman, David Gerard, DavidLevinson, Deltabeignet, Dutchy85, Easchiff, Eclecticology, Emperorbma, Fayenatic london, Feitclub, GorgeCustersSabre, Hans yulun lai, Ic0n0klast, Ishikawa Minoru, JGKlein, JSydel, Juliancolton, Kakaze, Kamkek, Karin127, KnightRider, Kusunose, LGagnon, LilHelpa, Lugnuts, Lusample,

MachoCarioca, ManymerrymenmakingmuchmoneyinthemonthofMay, Mel Etitis, Morio, Naniwako, Nice poa, Oxyqwerty, Paul Richter, Pegship, Phbasketball6, QuizzicalBee, Red Deer, Rich Farmbrough, Robertvan1, Scieberking, Sekicho, ShelfSkewed, Shoeofdeath, Sholom, SimonP, Skymasterson, Snek01, Sreejithk2000, Swimdb, Taipansteel, Ted Wilkes, Tedius Zanarukando, The smilodon, Thefourdotelipsis, Tientao, Tijfo098, Tjmayerinsf, Treybien, Underwaterbuffalo, Warpozio, Welsh, WhisperToMe, Wool Mintons, XinJeisan, Zoe, 36 anonymous edits

Miyoshi_Umeki *Source*: http://en.wikipedia.org/w/index.php?title=Miyoshi_Umeki *Contributors*: Adamthemillionth, Ameliorate!, Anthony Winward, Astorknlam, Badagnani, Bendono, Best name, Billy Hathorn, Black-Velvet, Bovineboy2008, Caeruleancentaur, Caldwell4, Canadian Paul, CanisRufus, CatchyJ, Cburnett, D6, DA19, Dekimasu, Drbreznjev, Ed Poor, Eliyahu S, Evan1975, Ewulp, ForDorothy, Fourthords, Garion96, Gcstackmoney, Gmosaki, GoingBatty, Good Olfactory, Gr8lyknow, Hohum3, IceBrotherhood, Italianlover07, JackofOz, JamesAM, Jennes83, Jetman, Joey80, Jrla05, JustAGal, Kbdank71, Kentucho, Krakatoa, Kumioko (renamed), Lucobrat, Lugnuts, MRH704, Mateusz Opasiński, MisfitToys, Mllefifi, Monkeyzpop, Mushroom, MusicMaker5376, Myasuda, Nicke Lilltroll, NoseNuggets, Pinkadelica, Pizzamaniac09, Professor Von Pie, Regent of the Seatopians, Rettetast, Rms125a@hotmail.com, Rossrs, RyanGerbil10, Scanlan, Ser Amantio di Nicolao, ShelfSkewed, Snek01, Srbauer, Ssilvers, Stoogetins, Tassedethe, Thismightbezach, Tientao, Time and again, Tinton5, Unclemikejb, Veggiegirl, We hope, WhisperToMe, Wildhartlivie, Wlparadise, Xophist, Yas, Zzyzx11, 111 anonymous edits

Hatari! *Source*: http://en.wikipedia.org/w/index.php?title=Hatari%21 *Contributors*: 2620:106:6000:20:0:0:0:39, Alifazal, Andrzejbanas, Apoc2400, Batdoguk, Belovedfreak, Bevo, Bobber100, BuddylBat, Cbrown1023, Chicnomad, Chris fardon, Cliff1911, Coelacan, Cooksey, CyberSkull, Cyberose, D6, DavidOaks, Deor, Discopriest, Dutchy85, Dylanfromthenorth, EamonnPKeane, FlashSheridan, Foofbun, Fuhghettaboutit, Grammar14, Harro5, Jbrianjones, Jimday172, John K, Kuralyov, Lockley, Lugnuts, MachoCarioca, Malangali, MarcusBritish, Monegasque, Muntuwandi, Neurolysis, Nmojtahed@msn.com, Polisher of Cobwebs, Prekario, Quentin X, Rebrane, RicJac, Rich Farmbrough, Russell ARKINSTALL, Savolya, Sreejithk2000, Stepho-wrs, Stygian23, The Founders Intent, The wub, Tucci78, Varlaam, Wastetimer, Ziggurat, 60 anonymous edits

The Longest Day (film) *Source*: http://en.wikipedia.org/w/index.php?title=The_Longest_Day_%28film%29 *Contributors*: 02nelli, A. Carty, A.R., Ab.chaplin, Accubam, Aj4444, AlbertSM, Andres, Andrzejbanas, Anthony Appleyard, Anthonywoodwark, Arakunem, AshLin, Beanbatch, Before My Ken, Bender235, Best O Fortuna, Beyond My Ken, Billy Hathorn, Bilsonius, Black Mesa, Blair Bonnett, BlueSalo, Bobet, Brookesward, CGameProgrammer, Cburnett, Charlie52, Chaser, Chris fardon, Colin marks, Cool3, Crotchety Old Man, DJ Clayworth, Dahveed76, Darev, Dasuitekilla, Dawkeye, Dbromage, Delldot, Derfel73, Destry, Dhartung, Dodgerblue777, Donmike10, Dr. Blofeld, Drogo Underburrow, Dutchy85, Ekwos, Elendil's Heir, Entrophie, Ericoides, Evil Monkey, Fah-Q 2, Films addicted, Florwood, Foofbun, Frecklefoot, Freshh, Fufu60601, Gael13011, Gary E Nelson, GeoffreyVS, Gerbrant, Gooberliberation, Gothicfilm, GraemeLeggett, Greenshed, GusF, Hamish59, Hexmaster, Hugo999, Infrogmation, J04n, JEklöw, JackofOz, JakobVX, Jay-W, Jeffrey O. Gustafson, Jeop86, Jim Sweeney, Jogers, John N., Johns78, Jooler, Jros83, Jsonitsac, K1Bond007, KConWiki, KD5TVI, Kalmbach, Kamkek, Karl eichholtz 13, Keallu, Keegan, Kelisi, Kernitou, Klemen Kocjancic, Konczewski, Ktr101, LB-Versender, Leflyman, Leoni2, Leothed, Lepeu1999, Londonclanger, Lord Cornwallis, LtNOWIS, Lugnuts, Lyricmac, MK2, MachoCarioca, Magnet For Knowledge, MarcusBritish, Marrante, Martin Rizzo, Member, Michael Dorosh, Millbart, Miraclewhip, Neddyseagoon, Neon white, NewEnglandYankee, Ng.j, Nitraven, Noirish, Nymf, Odder, Olivier, Oneiros, Openskye, Orbicle, Paste, Paul Barlow, Pere Serafi, Phbasketball6, Plasticspork, Prlambert76, QuiteUnusual, R. fiend, RP459, Rama, Rameriz, Reedmalloy, Renegade Replicant, RobNS, Rogerd, Rwils, RxS, Savolya, Scorpion0422, Sedelmayer, Ser Amantio di Nicolao, Shan246, Sherurcij, SidP, SigPig, Silverhorse, SimonATL, Sjyglm, Sknayfan, Skymasterson, SoLando, Sreejithk2000, Stackja1945, SteveCrook, Storyliner, Tedman1973, TeleComNasSprVen, Terraflorin, The Mercenary 73, The Saturday Boy, The wub, TheMovieBuff, TheOldJacobite, Thismightbezach, Thomas Blomberg, Thomaskorp, Tiddly Tom, Upsmiler, Ustye, Valentinejoesmith, Valentinian, Vaquero100, Warfreak, Wasted Time R, WikiDon, Wikijsmak, Will Beback, Woohookitty, Writerguy, Wwoods, Xezbeth, YUL89YYZ, Yankee961, Zapyon, Zoicon5, 292 anonymous edits

The Eleventh Hour (1962 TV series) *Source*: http://en.wikipedia.org/w/index.php?title=The_Eleventh_Hour_%281962_TV_series%29 *Contributors*: After Midnight, Azumanga1, Bearcat, BigT2006, Billy Hathorn, Chris the speller, Colonies Chris, Firsfron, Foofbun, Gaius Cornelius, GoingBatty, King Shadeed, Klemen Kocjancic, Lightmouse, Pepso2, Pinkadelica, Ser Amantio di Nicolao, Sky Captain, Tassedethe, Whytk, 2 anonymous edits

Little House on the Prairie (TV series) *Source*: http://en.wikipedia.org/w/index.php?title=Little_House_on_the_Prairie_%28TV_series%29 *Contributors*: 97198, Aaron Walden, AaronCBurke, Absinthe88, Adrruiz, Aldengirl, Allnewramki, Alrofficial, Ancjr, AnnieGrint91, Arthena, Aschrage, Auntof6, Azumanga1, Babygator23, Beaner142, Bearcat, Beetfarm Louie, Beetstra, Beland, Benedict J S, Billy Hathorn, Bmarq, Boing! said Zebedee, Bongwarrior, Bouaketh, Bovineboy2008, BrainyBabe, Briantw, Briguy52748, Calabe1992, ChrisJsc, Christopher Parham, Civil Engineer III, Ckatz, Clerks, CobraWiki, Codyfinke6, Colombe7138, Colonies Chris, Copana2002, Countrygirls199822, Craig Adams, Crakkerjakk, Creativity97, CrestwoodRocks, Cutter, Czolgolz, DHN, DVdm, Dale Arnett, DandyDan2007, David Gerard, DavidDouthitt, Davodd, Dell9300, Desertmonk, Dismas, Doogie1953, Dvd-junkie, Dwanyewest, Elliskev, Enviroboy, Erik9, ExRat, Extra mile 4 you, FMAFan1990, Fhb3, Finkefamily, Fireseeker4him, Freshh, Frongle, Gaius Cornelius, Gcollinsii, Gcstackmoney, GeneralBelly, Gilliam, Goldenroad, GregorB, GrimGrinningGuest, Gwendibbley, Haleemah Shah, Hans100, Hard Luck Harry, Hendumaica, Hiphats, Hossain Akhtar Chowdhury, Hrsn, Hugo999, Iam4Lost, Ignotus91987, Imroy, Ipodlvr124, Iridescent, Italianlover07, IzzysGirl, JAYMEDINC, JMyrleFuller, JamesBWatson, JanisLyn, Jason Palpatine, JasonAQuest, Jbl1975, Jc37, Jessica93, Jim856796, JimBarbasol, John Goemley, Jojhutton, KGasso, Kane5187, Kangaroo guy, Khairul hazim, Khazar2, King Shadeed, Kinston eagle, Kitch, Kyleall, Kyros, Lafong, Larry hime, Leepetmar, LeoStarDragon1, Lotje, Loudounlib, LucyDoo, MDCollins, Mak1457, MakeRocketGoNow, Materialscientist, Matthew, Mehdi, Mike Halterman, Milkmeister, Mjrmtg, Moncrief, Monkey Bounce, Mrschimpf, Muppets101, Musdan77, MusicalTheatreGirl, Ncmattj, Neelix, Niki2006, Nodekeeper, Odikuas, Ohconfucius, Ommnomnomgulp, Overkill82, PCHS-NJROTC, Pastor Kam, Paulmcdonald, Pennydarwin, PhnomPencil, Pinkadelica, Pizzamaniac09, Postcard Cathy, Postdlf, Prof .Woodruff, Pstril, Pwolf226, R7604, RFD, Random86, Rayburne1997, Red Director, RobertCMWV1974, Robinepowell, Rogerd, Rolypolyman, Rypcord, Shannernanner, Sjsitler92, Skywalker80100, Sngnisfuk, SonPraises, Stbalbach, Stephen Bain, Sterling Holobyte, Stusutcliffe, Supursillysally, TPIRFanSteve, TVDB, Tedder, Telefan, The Nuke, The27thMaine, Thivierr, Thor Dockweiler, Timvasquez, Tommy2010, TonyTheTiger, Tregoweth, Trezjr, TyA, Unintentional Guy, Updatehelper, Vanished user 39948282, Vuono, Wayne Slam, We hope, Wesley, Wheeler0152, Wikijmt, Wildhartlivie, Wlmg, Woohookitty, Wwoods, Xiahou, Yamla, Ynhockey, You're Doing Well, Younome, Zooks527, Zzyzx11 the clone, 456 anonymous edits

It's Garry Shandling's Show *Source*: http://en.wikipedia.org/w/index.php?title=It%27s_Garry_Shandling%27s_Show *Contributors*: 23skidoo, Benscripps, Bigoranget, Bookemdano63, Branddobbe, Cab88, Caldorwards4, CobraWiki, D0762, Damiwh2, Davidhorman, Deliamber, Ds13, Dwanyewest, Elassint, FMAFan1990, Fernandobouregard, Fred J, Frietjes, GrahamHardy, Holden190, Huh direction, Iam4Lost, Idols of Mud, Iohannes Animosus, JamesAM, JeLuF, Jlouise311, Jmaxwell2400, Joedeshon, Jschuur, Karl-Henner, Lambertman, MK, MakeRocketGoNow, Martarius, Meco, Michael E Lynch, Mordea, NuclearWarfare, Pegship, Pinkadelica, R.A Huston, Rayc, RedWolf, Retropolis1, Rjwilmsi, Rlquall, Robert Moore, Rogerd, Satori Son, ShelfSkewed, SidP, Skywalker80100, StewieBaby05, TMC1982, TPIRFanSteve, The Anome, The GateKeeper07, The JPS, Theresa knott, ThunderPeel2001, Tony1, TonyTheTiger, TwoTwoHello, VirtualDelight, Woohookitty, Wrightaway, Zpb52, 68 anonymous edits

Comedy_Central *Source*: http://en.wikipedia.org/w/index.php?title=Comedy_Central *Contributors*: 0v1d1o114, 10Rooneyfan10, 123abc345def, 1836311903, 24 biggest fan, 28421u2232nfenfcenc, 53180, A895, AAA!, APT41790, AT3397, Acroterion, Acsenray, Actual User, After Midnight, Aitias, Ak3883, Al Wiseman, Alansohn, Aldog2k12, Alex43223, Aliquidparadigm, Alphaballs, Amiodarone, Anangryblackperson, Andrewlp1991, Android79, Andy Marchbanks, Angie Y., Ank329, Anoninmous, Antandrus, AnyPerson, Aots191, Apparition11, Arthena, Ascidian, AshTFrankFurter2, Ashburn247, Aspects, Assassin.gonzalez01, Astavats, Ausir, Avicennasis, AxelBoldt, Azumanga1, BHayes07, BakerKr3w, Bassbonerocks, Bbb2007, Bearcat, BeefCake, Bejinhan, Ben Ben, Bencossette, Benscripps, Benzy19, Bfisd, Bflorio12, Bhall87, BigJolly9, Biggumsmacarthur, Billy Liakopoulos, Bjones, Blackholes11, BlairsvilleHighSchool, Blasberuf, Blassj, Blerickjong, BlueMario1016, Bmitchelf, Bobak, Bobbiclayton, Bobblehead, Bobrayner, Bockandrock, Boisemedia, BonesBrigade, Bongwarrior, Bonifier, BookDen, Boomshadow, Boredgod, BornonJune8, Bradeos Graphon, Brainulator9, Branddobbe, Brandon5485, Briantw, Bull-Doser, Buttreyfood, CR85747, Calaschysm, Caldorwards4, CaribDigita, CastAStone, Cburnett, Ccm043, CharacterZero, Chester320, ChesterG, Chicagos finest, Chickenmonkey, Chief0224, Chill Pill Bill, Chkurz, Chris the speller, ChrisBradley, Christian75, Cielomobile, Claidheamohmor, Clayburn, Clegacyl, Clemson12378, Clubsammich, Coneslayer, Conner111213, Corn cheese, Courcelles, Cpatt18, Crash Underride, Crazytonyi, Cyberevil, CynicalRealist, Cyzor, D.scain.farenzena, DMVillan, DStoykov, DT29, Dabomb87, Dahveed76, Dalkaen, Damian1234567, Damienchock, Dcljr, Ddespie@san.rr.com, Dearsergiosilence, Deathawk, DelTribe, DeltaQuad, DesertEagle73, Doc Strange, DocWatson42, Dp76764, Drew31596, Drod98, Dstebbins, Dvda92, Dycedarg, EHRLT, ESkog, EagerToddler39, Edenc1, EdwinHJ, Egrian, Eisnel, ElSaxo, EmiOfBrie, Empoor, ErikNY, Escape Orbit, Estoy Aquí, Evan916, EvanOven12, Everyking, Evice, Evildevil, Excirial, Expertspecter, FMAFan1990, Faded, Fairlyoddparents1234, Fatbusinessman, Fernandobouregard, Flavoice, Fleutist, FluffyIndieKid, Flyguy649, Forteblast, Friginator, Funnyfarmofdoom, Fuzheado, Fuzzy901, GVOLTT, Gadfium, Gamingman34, Gareth Griffith-Jones, Gasoffer, George Adam Horváth, Gfoley4, Gigitel, Gkjsamr, Glass Sword, Glassbreaker5791, Glen, GoShow, Gobonobo, Gogo Dodo, Gpomeroy95, Grammarxxx, Grapesoda22, Gremo32, Grfxfreak, Grindor, Gtrmp, H2g2bob, HPJoker, Hacker2000, Haham hanuka, Hailey C. Shannon, HalJor, Haleykern19, HallucigeniaUK, HangingCurve, HarlandQPitt, Hawaiian717, HeidiFloss, Helloitz43, Heroville, HinoKagu, Hiphats, Hmr, Hoponpop69, Hurgeo2320, Hypernick1980, I like fun, IAmCarlosMencia, IDALGHAMTFPD, ISD, Iam4Lost, Ike9898, Ilpfilms, Imorse98, Inordinate, Insanity Incarnate, Inventm, IrisKawling, Irish Souffle, IronGargoyle, Ithizar, J. Nguyen, J.delanoy, JB82, JDoorjam, JKBrooks85, Jack's Revenge, Jackass2009, JakeDHS07, JamesReyes, Jason One, Jason.cinema, Jatkins, Jay Bones, Jayrod0123, Jc126fda, Jcbarr, Jeffq, Jehochman, Jimjones192, Jj98, Jnorton7558, Joeyhens, Joezone12, Jog13, John, John Pannozzi, John123521, JohnCub, Joltman, Jordao095, Joseph Solis in Australia, Josh3736, Josio00, Jvbishop, Jvhertum, Jéské Couriano, Kabumaru, Kafziel, Kaldari, Kaleb669, Karthikndr, Kateshortforbob, Kazvorpal, Kejoxen, Ken Watanabe, Kenny 2 da max, Kh80, King scooter 123, Krumley2007, Kubigula, Kundor, Kurieeto, KyraVixen, L Kensington, Lachatdelarue, Lahiru k, Lake9821, Larsenat, Latics, Laugie.0, Lee Cremeans, LeoNomis, Levineps, Lexein, Liface, Lightmouse, Litalex, Little Mountain 5, Little green rosetta, Llec, Lucamryn, Ludawest, Lukeyjay, Luna Santin, MAKAVELI 87, Maatc, Mach240390, Magical hand, Majkul, Marcberm, MarcelZimmermann, Marek69, Marfan8, Mariacer Cervantes, MarkSweep, MarkThreesixty, MarsNova, Martarius, Masman10, Matt319, Mauler90, MaxPride, Mbmac01, Mbrstooge, McSly, Mcfly85, Mcoop06, Meathead76, Meco, Meelar, Melvinsneedly, Mermaid from the Baltic Sea, Methodman321, Mhking, Michael Snow, Michaelas10, MickMacNee, Mikepjones, Minna Sora no Shita, Miranda, Mirv, Misternuvistor, Misterrick, Mitcheezy, Mogism, Mole, MonsterMediaQC, Montaced, Moomoo24, Mqduck, Mrblondnyc, Mrschimpf, Msolh1984, Mtjaws, Mulad, Murph3pt, Mushroom9, MusicInTheHouse, Mvtim3, NYKevin, Nard the Bard, NedBigby3, NekoDaemon, Nickelodeonfan2007, Ninja200640, NintendoFan11, Nivix, Nixrsix, Nolessthanevil82, Northwest, Nua eire, Nv8200p, ONEder Boy, Obi-WanKenobi-2005, Obradovic Goran, Of, Ohconfucius, Olevipoeg, OliverDF, Omicronpersei8, Orangemike, Osopaja, Ottoflippa, PCHS-NJROTC, PM800, Paparoster, Patstuart, Pdcook, Peanutbutterstella, Pherdy, Pikolas, Pilotguy, Planet-man828, Polarbear97, Postdlf, Ppntori, Prede, PrincessofLlyr, ProbablyX, PsychoJason, PudNoob, Qqzzzz, Qwertyuiop16095, RA0808, REGNILLO, RUL3R, Radiojon, Ram3500, RaresPaul, Rayshon, Rd123, Reahad, RealTNAFan, Reaverdrop, Redfarmer, Redtheidiot, Rettetast, Reub2000, RevRaven, ReyRichard, Ricardojose20027, Riffsyphon1024, RoadRunning13, Roadrunner3000, Ron Ritzman, RoyBoy, Rrburke, Rumplemayer, Runescapeisverygay, Rypcord, SMC, ST47, Sailorptah, Salty1984,

Samgoly, SamuraiClinton, ScottMHoward, Scottcraig, Scoub, Scrabbleship, Sean Quixote, Sebandreas, Sethlutz895, Shanes, Shea DM, Shkarter1985, Shortride, Shsilver, SigKauffman, Silverfish, SimonArlott, SimonP, Sintonak.X, SirPeebles, Skeleton 777, Skysmurf, Sledgeh101, SluggoOne, Slysplace, Smashmasterfox, Smile a While, Smoth 007, Socks 01, SomeGuy11112, Someguy1221, SouthernMan, Sp!ke, SpaceFlight89, Sperril, Sphyrune, Spidlock, Spikey, Sriharsh1234, Stardust8212, Stevietheman, Steviethunders, Stickguy, Stitchon, Suffusion of Yellow, Sully, Sundeuce98, SuperDude115, Superman212, Surachit, Svgalbertian, Swellman, TAnthony, TBrandley, TITROTU, TJ Spyke, TMC1982, TPIRFanSteve, Tabletop, Tanaats, Tassedethe, Tbhotch, Tdubbs1234, Teamzambo, Teethmonkey, Th3 k3y mast3r, Thatother1dude, The Filmaker, The JPS, The Rogue Penguin, The sober irishman, The wub, TheCustomOfLife, TheGoofyGolfer, TheOldJacobite, TheREALCableGuy, Thebestofall007, Thegreatmonkey, Thexper, Think outside the box, Tiddly Tom, TigerK 69, Tim1357, Tinton5, Toad of Steel, Tocino, Tofutwitch11, Tom, TomCat4680, TomStike, Tomg91, Tomjerry9, Tommy11111, Tommy8330, Torontonian1, Tregoweth, Trjumpet, Trogga, Trollingshiz, TruPepitoM, Trust Is All You Need, Ttenchantr, Turk brown, Tv's emory, Tvtonightokc, Twsx, Uncle Dick, Unreal7, UpstateNYer, VIRUZ671, Valley2city, Vegaswikian, Velvetron, Vinty, ViperSnake151, Vishnu2011, Volker89, Vyran, Waggers, Wavy G, Wayland, Wcquidditch, Weatherman87, Whiteraven17, Wickedjacob, Wiendietry, Wiki libs, WikiDani, Wikiman21387, Windchaser, WizardDuck, Woohookitty, WorthWatching, Xezbeth, Xxvid, Yanksox, Yowiki, ZJP, ZKDot, ZPM, Zach425, Zer0721, Zidonuke, Zone46, Zoz, ZrzlKing, Ztrewq321, Zzuuzz, شرزر, 927 anonymous edits

Knots Landing *Source*: http://en.wikipedia.org/w/index.php?title=Knots_Landing *Contributors*: 23skidoo, 97198, ABCxyz, AdamDeanHall, Adamwade, After Midnight, AlbertGray, AlexDitto, Alrofficial, Amalas, Andrei Cvhdsee Brazil, Angusmclellan, Anotherwikipedian, Aphaia, Ardonik, Arjayay, Artemisboy, Asa01, AuburnPilot, AvicAWB, AxG, BAYCITY99, BRDave01, Badagnani, Bearcat, BeckiGreen, Ben King, Ben-w, Billy Hathorn, Blake Burba, Bobo192, Bodil, Booshakla, BornonJune8, Bryancrowusa, Bryanrcrow, Calista06, CaseKid, Chris the speller, Cmdrjameson, Cmr08, CobraWiki, Colonies Chris, CowboySpartan, Crazyseiko, Damon207, Darinaria, David Gerard, Davidw0998, Dawynn, Dayewalker, Dbart, Dma124, Dmn, Dowew, DoxTxob, Drpickem, Dthomsen8, DtownG, Dugwiki, Dwanyewest, Econrad, Edward, Erasmussen, Esmeralda10, Eugene-elgato, Firestorm566, Firsfron, Freefry, Funkeekatt, Furrykef, Gcollinsii, Gcollinsii@aol.com, Godaiger, GoingBatty, GregorB, GrimGrinningGuest, Hadal, Hans100, Hardnfast, HatsuBeikokusekiryuso, Hebrides, Heegoop, Hiphats, HorrorFan121, Iam4Lost, Ian Pitchford, Infousabrc, Italianlover07, Jagarin, Jaldridge86, JamesB3, JamesTeterenko, Jessejames412, Jim856796, John of Reading, JulianthePhilosopher, Juppiter, JustAGal, Kaisershatner, Kbthompson, Kefas.se, KevinKale, Keycard, Khazar, Knowledgeman800, Kythena, Ladydayelle, LarRan, Lexdaleowens, Liface, LilHelpa, Lordgilman, Lowellian, Luvhart, Lvr, MHarrington, MK2, Mahanga, MakeRocketGoNow, Mandarax, Markeer, Marmite Man, MarnetteD, MassassiUK, Mattbr, Matthew, MegX, Mellery, Melrose485, Mike Halterman, Mistahollywood, Msclguru, Nakon, Neilc, Night Tracks, No1lakersfan, Northofdc, Oatmeal batman, Otto4711, Paulmcdonald, Pinkadelica, Pizzamaniac09, Quadell, Queenmomcat, RadioKirk, Radiohawk, Rassatee, RedWolf, Rjwilmsi, Rlquall, RobBoston, Robert Moore, RobertCMWV1974, Robinnfletcher, Rphr1987, Ruby Bartlett, Ryulong, SD6-Agent, SFTVLGUY2, Sam Hocevar, Scarian, Scherezade, SchuminWeb, Scorpion0422, Seansinc, Sfjpk30, Shcuts, ShelfSkewed, Shkarter1985, SiobhanHansa, Something X, Sschlimgen, TAnthony, TMC1982, TTN, Tassedethe, Ted Wilkes, Thames5, Thatssomike, The Flagpole, The JPS, The stuart, TheCustomOfLife, TheHYPO, Tinmanic, Tom Morris, Tony1, Trekphiler, Trueslpatriot, Uide, Veesicle, Vnisanian2001, Vonvon, Wabbit98, Wac01, WereSpielChequers, Who, Whoville, Wwoods, Wyn9, Yannick41, Ynhockey, ZillowEdge, ZouBEini, 521 anonymous edits

The Dean Martin Celebrity Roast *Source*: http://en.wikipedia.org/w/index.php?title=The_Dean_Martin_Celebrity_Roast *Contributors*: Artistjakefoster, Ben76266, Bt8257, Cliff1911, Cush, Demomoke, JMyrleFuller, Jacroe, JustAGal, Ketiltrout, LittleWink, Lockley, MaJic, MegastarLV, Noozgroop, PKT, RHodnett, SWLinPHX, Stundra, Tassedethe, Tinton5, Vegaswikian1, Walor, 41 anonymous edits

Century_Village,_Florida *Source*: http://en.wikipedia.org/w/index.php?title=Century_Village%2C_Florida *Contributors*: Andrwsc, Averette, DavidLevinson, Discospinster, Donald Albury, Edward, Giraffedata, Goustien, Jeffcpix, John Cardinal, Jorfer, LCU03, OxyMoronMinusOxy, Pearle, Ram-Man, Seth Ilys, Twalls, Vegaswikian, 6 anonymous edits

Anna_Kashfi *Source*: http://en.wikipedia.org/w/index.php?title=Anna_Kashfi *Contributors*: 81 with love, All Hallow's Wraith, Appraiser, Bearcat, Benjamin23 allen, Boguslinks, CecilF, Chuck56, Clarityfiend, Corstarl, CruftEater, D6, Ekabhishek, George Ho, GoingBatty, Gustav von Humpelschmumpel, JackofOz, Kathanar, Kintetsubuffalo, Kitchawan, Legalwatchdog, Lightmouse, Lugnuts, Orbicle, Pixelface, Princesshapnick, RLLsecondcorp, Rich Farmbrough, Rms125a@hotmail.com, Rrsmac, Selbymayfair, Supertigerman, TheOldJacobite, Tjmayerinsf, Werdnawerdna, Wheatfields7, Yamanbaiia, 27 anonymous edits

Movita_Castaneda *Source*: http://en.wikipedia.org/w/index.php?title=Movita_Castaneda *Contributors*: All Hallow's Wraith, Arniep, Bearcat, BrownHairedGirl, Chicheley, Coolavokig, Deanlaw, GusF, Gustav von Humpelschmumpel, ISAYsorry, Jennica, Johnpacklambert, Mocu, Philip Cross, Plasticspork, Ron whisky, Rossrs, SamChambers, Savolya, Shining.Star, TexasAndroid, Ustye, Walter Breitzke, Wikid77, 18 anonymous edits

Tarita_Teriipia *Source*: http://en.wikipedia.org/w/index.php?title=Tarita_Teriipia *Contributors*: AQUIMISMO, Anthony Winward, Arniep, Cherkash, Clarityfiend, Doncsecz, Fvw, GlassCobra, Good Olfactory, Grutness, Gustav von Humpelschmumpel, HerbertMMarx, Honeycake, Ionesco, JackofOz, Kjkolb, Magioladitis, Maralia, Mike Selinker, Mythic Writerlord, Obli, Polylerus, Rbraunwa, Rbrwr, Samantha of Cardyke, Savolya, Son of Somebody, Steven Zhang, SwanSZ, Tamariki, TheMadBaron, TheParanoidOne, Waacstats, Walter Breitzke, Wizardman, Yemal, Zoicon5, 16 anonymous edits

Image Sources, Licenses and Contributors

File:RedButtonsAL1978.jpg *Source*: http://en.wikipedia.org/w/index.php?title=File:RedButtonsAL1978.jpg *License*: unknown *Contributors*: Alan Light at http://www.flickr.com/photos/alan-light/

File:Red Buttons Double Life of Henry Phyfe 1965.JPG *Source*: http://en.wikipedia.org/w/index.php?title=File:Red_Buttons_Double_Life_of_Henry_Phyfe_1965.JPG *License*: unknown *Contributors*: ABC Television

File:Gnome globe current event.svg *Source*: http://en.wikipedia.org/w/index.php?title=File:Gnome_globe_current_event.svg *License*: unknown *Contributors*: User:Anomie, User:Davidgothberg

File:Oscar statuette.jpg *Source*: http://en.wikipedia.org/w/index.php?title=File:Oscar_statuette.jpg *License*: unknown *Contributors*: BomBom, Rzuwig, Skier Dude, Tbhotch, Zain.3nov, Zzyzx11, 2 anonymous edits

File:FONTAINE-COOPER.jpg *Source*: http://en.wikipedia.org/w/index.php?title=File:FONTAINE-COOPER.jpg *License*: unknown *Contributors*: Darekm135, Magog the Ogre, Snorri95

File:31st Acad Awards.jpg *Source*: http://en.wikipedia.org/w/index.php?title=File:31st_Acad_Awards.jpg *License*: unknown *Contributors*: See above

File:81st Academy Awards Ceremony.JPG *Source*: http://en.wikipedia.org/w/index.php?title=File:81st_Academy_Awards_Ceremony.JPG *License*: unknown *Contributors*: BDS2006 (talk)

File:Hollywood Pantages Theatre 5.jpg *Source*: http://en.wikipedia.org/w/index.php?title=File:Hollywood_Pantages_Theatre_5.jpg *License*: unknown *Contributors*: Photo: Andreas Praefcke

File:Fiorello LaGuardia Signature.svg *Source*: http://en.wikipedia.org/w/index.php?title=File:Fiorello_LaGuardia_Signature.svg *License*: unknown *Contributors*: Fiorella LaGuardia

Image:F.H.LaGuardia in front CaproniCa.44 bomber.jpg *Source*: http://en.wikipedia.org/w/index.php?title=File:F.H.LaGuardia_in_front_CaproniCa.44_bomber.jpg *License*: unknown *Contributors*: Il palazzo, Infrogmation, Threecharlie

Image:LaGuardia In.JPG *Source*: http://en.wikipedia.org/w/index.php?title=File:LaGuardia_In.JPG *License*: unknown *Contributors*: User:Smarticvs

File:LaGuardia Congress.jpg *Source*: http://en.wikipedia.org/w/index.php?title=File:LaGuardia_Congress.jpg *License*: unknown *Contributors*: Collection of U.S. House of Representatives

Image:Fiorella LaGuardia statue.jpg *Source*: http://en.wikipedia.org/w/index.php?title=File:Fiorella_LaGuardia_statue.jpg *License*: unknown *Contributors*: User:Mskyrider

Image:Fiorello LaGuardia Grave 1024.jpg *Source*: http://en.wikipedia.org/w/index.php?title=File:Fiorello_LaGuardia_Grave_1024.jpg *License*: unknown *Contributors*: Anthony22 (talk). Original uploader was Anthony22 at en.wikipedia

File:Fiorello LaGuardia footstone 800.jpg *Source*: http://en.wikipedia.org/w/index.php?title=File:Fiorello_LaGuardia_footstone_800.jpg *License*: unknown *Contributors*:

File:Robeson Hagen Othello.jpg *Source*: http://en.wikipedia.org/w/index.php?title=File:Robeson_Hagen_Othello.jpg *License*: unknown *Contributors*: DionysosProteus, FSII, Infrogmation

image:Lanza still MGM Cesari.jpg *Source*: http://en.wikipedia.org/w/index.php?title=File:Lanza_still_MGM_Cesari.jpg *License*: unknown *Contributors*: MGM studios - see Talk for more details

Image:Mario Lanza Otello.jpg *Source*: http://en.wikipedia.org/w/index.php?title=File:Mario_Lanza_Otello.jpg *License*: unknown *Contributors*: MGM

Image:Tucker Lanza.jpg *Source*: http://en.wikipedia.org/w/index.php?title=File:Tucker_Lanza.jpg *License*: unknown *Contributors*: User:NewYork1956

Image:Mario Lanza Pinkerton.jpg *Source*: http://en.wikipedia.org/w/index.php?title=File:Mario_Lanza_Pinkerton.jpg *License*: unknown *Contributors*: MGM

File:John Forsythe in The Trouble With Harry trailer.jpg *Source*: http://en.wikipedia.org/w/index.php?title=File:John_Forsythe_in_The_Trouble_With_Harry_trailer.jpg *License*: unknown *Contributors*: Trailer screenshot

Image:Forsythe.JPG *Source*: http://en.wikipedia.org/w/index.php?title=File:Forsythe.JPG *License*: unknown *Contributors*: Original uploader was Neelix at en.wikipedia

File:John Forsythe star HWF.JPG *Source*: http://en.wikipedia.org/w/index.php?title=File:John_Forsythe_star_HWF.JPG *License*: unknown *Contributors*: User:JGKlein

File:Karl Malden 2.jpg *Source*: http://en.wikipedia.org/w/index.php?title=File:Karl_Malden_2.jpg *License*: unknown *Contributors*: Karl's friend Barry Iddon, Variety Club of Great Britain

File:Karl Malden's grave at Westwood Village Memorial Park Cemetery in Brentwood, California - December 2011.jpg *Source*: http://en.wikipedia.org/w/index.php?title=File:Karl_Malden's_grave_at_Westwood_Village_Memorial_Park_Cemetery_in_Brentwood,_California_-_December_2011.jpg *License*: unknown *Contributors*: User:Meribona

File:Karl Malden in I Confess trailer.jpg *Source*: http://en.wikipedia.org/w/index.php?title=File:Karl_Malden_in_I_Confess_trailer.jpg *License*: unknown *Contributors*: Trailer screenshot

File:Karl malden marlon brando waterfront 4.jpg *Source*: http://en.wikipedia.org/w/index.php?title=File:Karl_malden_marlon_brando_waterfront_4.jpg *License*: unknown *Contributors*: Trailer screenshot

File:Karl malden eva marie saint waterfront 1.jpg *Source*: http://en.wikipedia.org/w/index.php?title=File:Karl_malden_eva_marie_saint_waterfront_1.jpg *License*: unknown *Contributors*: Trailer screenshot

File:Karl malden on the waterfront 2.jpg *Source*: http://en.wikipedia.org/w/index.php?title=File:Karl_malden_on_the_waterfront_2.jpg *License*: unknown *Contributors*: Trailer screenshot

File:Lee J. Cobb 1960s.JPG *Source*: http://en.wikipedia.org/w/index.php?title=File:Lee_J._Cobb_1960s.JPG *License*: unknown *Contributors*: NBC Television

File:William Holden-Cobb-Golden Boy.jpg *Source*: http://en.wikipedia.org/w/index.php?title=File:William_Holden-Cobb-Golden_Boy.jpg *License*: unknown *Contributors*: Studio

File:Lee j cobb brando on the waterfront 2.jpg *Source*: http://en.wikipedia.org/w/index.php?title=File:Lee_j_cobb_brando_on_the_waterfront_2.jpg *License*: unknown *Contributors*: Trailer screenshot

File:Lee j cobb on the waterfront 4.jpg *Source*: http://en.wikipedia.org/w/index.php?title=File:Lee_j_cobb_on_the_waterfront_4.jpg *License*: unknown *Contributors*: Trailer screenshot

File:Darryl F. Zanuck in Grapes of Wrath trailer.jpg *Source*: http://en.wikipedia.org/w/index.php?title=File:Darryl_F._Zanuck_in_Grapes_of_Wrath_trailer.jpg *License*: unknown *Contributors*: Trailer screenshotLicencing information :http://www.sabucat.com/?pg=copyright and http://www.creativeclearance.com/guidelines.html#D2

File:Darryl Zanuck grave at Westwood Village Memorial Park Cemetery in Brentwood, California.JPG *Source*: http://en.wikipedia.org/w/index.php?title=File:Darryl_Zanuck_grave_at_Westwood_Village_Memorial_Park_Cemetery_in_Brentwood,_California.JPG *License*: unknown *Contributors*: User:Meribona

file:Winged_Victory(film).jpg *Source*: http://en.wikipedia.org/w/index.php?title=File:Winged_Victory(film).jpg *License*: unknown *Contributors*: AvicAWB, Jbmas99, Salavat

File:George Cukor Allan Warren.jpg *Source*: http://en.wikipedia.org/w/index.php?title=File:George_Cukor_Allan_Warren.jpg *License*: unknown *Contributors*: User:Allan warren

File:George Cukor 5 Allan Warren.jpg *Source*: http://en.wikipedia.org/w/index.php?title=File:George_Cukor_5_Allan_Warren.jpg *License*: unknown *Contributors*: User:Allan warren

File:George Cukor garden.jpg *Source*: http://en.wikipedia.org/w/index.php?title=File:George_Cukor_garden.jpg *License*: unknown *Contributors*: User:Allan warren

File:Second world war europe animation small.gif *Source*: http://en.wikipedia.org/w/index.php?title=File:Second_world_war_europe_animation_small.gif *License*: unknown *Contributors*: User:San Jose

File:Hitlermusso2 edit.jpg *Source*: http://en.wikipedia.org/w/index.php?title=File:Hitlermusso2_edit.jpg *License*: unknown *Contributors*: CillanXC, Excirial, F l a n k e r, Howcheng, J Milburn, JGHowes, Jbarta, MER-C, Mindmatrix, Monty845, Zzyzx11, 8 anonymous edits

File:Winter war.jpg *Source*: http://en.wikipedia.org/w/index.php?title=File:Winter_war.jpg *License*: unknown *Contributors*: Finnish Army personnel

File:Nazi-parading-in-elysian-fields-paris-desert-1940.png *Source*: http://en.wikipedia.org/w/index.php?title=File:Nazi-parading-in-elysian-fields-paris-desert-1940.png *License*: unknown *Contributors*: Frank Capra (director), U.S. War Department

File:Supermarinespitfire.JPG *Source*: http://en.wikipedia.org/w/index.php?title=File:Supermarinespitfire.JPG *License*: unknown *Contributors*: RAF official photographer

File:Yalta summit 1945 with Churchill, Roosevelt, Stalin.jpg *Source*: http://en.wikipedia.org/w/index.php?title=File:Yalta_summit_1945_with_Churchill,_Roosevelt,_Stalin.jpg *License*: unknown *Contributors*: Adam Zábranský, Butko, EurekaLott, Favonian, Florival fr, Gaaarg, Geni, George McFinnigan, Greenshed, Herbythyme, Jkelly, Mariluna, MarkSweep, Morio, Movieevery, Schaengel89, Shizhao, Spider death, Starscream, Surya Prakash.S.A., Thuresson, UKPhoenix79, Yann, 20 anonymous edits

File:De Gaulle-OWI.jpg *Source*: http://en.wikipedia.org/w/index.php?title=File:De_Gaulle-OWI.jpg *License*: unknown *Contributors*: Office of War Information, Overseas Picture Division. The image prefix (LC-USW3) at the Library of Congress image page matches that of pictures from the OWI collection (see prefix list here.

File:Churchill waves to crowds.jpg *Source*: http://en.wikipedia.org/w/index.php?title=File:Churchill_waves_to_crowds.jpg *License*: unknown *Contributors*: User:W.wolny

File:1943-07-01GerWW2BattlefrontAtlas.jpg *Source*: http://en.wikipedia.org/w/index.php?title=File:1943-07-01GerWW2BattlefrontAtlas.jpg *License*: unknown *Contributors*: Army Map Service

File:1943-11-01GerWW2BattlefrontAtlas.jpg *Source*: http://en.wikipedia.org/w/index.php?title=File:1943-11-01GerWW2BattlefrontAtlas.jpg *License*: unknown *Contributors*: Army Map Service

File:1944-07-01GerWW2BattlefrontAtlas.jpg *Source*: http://en.wikipedia.org/w/index.php?title=File:1944-07-01GerWW2BattlefrontAtlas.jpg *License*: unknown *Contributors*: Army Map Service

File:1944-09-01GerWW2BattlefrontAtlas.jpg *Source*: http://en.wikipedia.org/w/index.php?title=File:1944-09-01GerWW2BattlefrontAtlas.jpg *License*: unknown *Contributors*: Army Map Service

File:1944-12-15GerWW2BattlefrontAtlas.jpg *Source*: http://en.wikipedia.org/w/index.php?title=File:1944-12-15GerWW2BattlefrontAtlas.jpg *License*: unknown *Contributors*: Army Map Service

File:1945-05-01GerWW2BattlefrontAtlas.jpg *Source*: http://en.wikipedia.org/w/index.php?title=File:1945-05-01GerWW2BattlefrontAtlas.jpg *License*: unknown *Contributors*: Army Map Service

File:Mickey Rooney still.jpg *Source*: http://en.wikipedia.org/w/index.php?title=File:Mickey_Rooney_still.jpg *License*: unknown *Contributors*: Studio publicity still. Original uploader was Wikiwatcher1 at en.wikipedia

File:Love Finds Andy Hardy trailer.JPG *Source*: http://en.wikipedia.org/w/index.php?title=File:Love_Finds_Andy_Hardy_trailer.JPG *License*: unknown *Contributors*: Trailer screenshot

File:Mickey Rooney ww2 46.jpg *Source*: http://en.wikipedia.org/w/index.php?title=File:Mickey_Rooney_ww2_46.jpg *License*: unknown *Contributors*: Ary29, Djembayz, Man vyi, PMG

File:Rooney Skelton 1962.JPG *Source*: http://en.wikipedia.org/w/index.php?title=File:Rooney_Skelton_1962.JPG *License*: unknown *Contributors*: User:SreeBot

File:Uso-show-mickey-rooney.jpg *Source*: http://en.wikipedia.org/w/index.php?title=File:Uso-show-mickey-rooney.jpg *License*: unknown *Contributors*: SGT. Joseph Tonetti, USMC.

File:Mickey Rooney 2 Allan Warren.jpg *Source*: http://en.wikipedia.org/w/index.php?title=File:Mickey_Rooney_2_Allan_Warren.jpg *License*: unknown *Contributors*: User:Allan warren

Image:Jazz ensemble - seating diagram.svg *Source*: http://en.wikipedia.org/w/index.php?title=File:Jazz_ensemble_-_seating_diagram.svg *License*: unknown *Contributors*: User:LtPowers

Image:Whitemanband1921.jpg *Source*: http://en.wikipedia.org/w/index.php?title=File:Whitemanband1921.jpg *License*: unknown *Contributors*: EugeneZelenko, FredrikT, Infrogmation, Magister Mathematicae

Image:BennyGoodmanandBandStageDoorCanteen.jpg *Source*: http://en.wikipedia.org/w/index.php?title=File:BennyGoodmanandBandStageDoorCanteen.jpg *License*: unknown *Contributors*: Film screenshot

Image:Glen miller.jpg *Source*: http://en.wikipedia.org/w/index.php?title=File:Glen_miller.jpg *License*: unknown *Contributors*: Original uploader was SeanO at en.wikipedia

Image:OckbrookBandPridePark.jpg *Source*: http://en.wikipedia.org/w/index.php?title=File:OckbrookBandPridePark.jpg *License*: unknown *Contributors*: Peter cowley

file:Original movie poster for the film Sayonara.jpg *Source*: http://en.wikipedia.org/w/index.php?title=File:Original_movie_poster_for_the_film_Sayonara.jpg *License*: unknown *Contributors*: Calmer Waters, King of the North East, MachoCarioca, TheDJ, Thomp

File:Miyoshi Umeki.jpg *Source*: http://en.wikipedia.org/w/index.php?title=File:Miyoshi_Umeki.jpg *License*: unknown *Contributors*: OceanSound, Reggaeman, WTCA

file:Hatari.jpg *Source*: http://en.wikipedia.org/w/index.php?title=File:Hatari.jpg *License*: unknown *Contributors*: King of the North East, MachoCarioca, Skier Dude

file:Original movie poster for the film The Longest Day.jpg *Source*: http://en.wikipedia.org/w/index.php?title=File:Original_movie_poster_for_the_film_The_Longest_Day.jpg *License*: unknown *Contributors*: Calmer Waters, Juliancolton, MachoCarioca, Oxymoron83, Skier Dude, Thomp, 2 anonymous edits

File:John Wayne in The Longest Day trailer.jpg *Source*: http://en.wikipedia.org/w/index.php?title=File:John_Wayne_in_The_Longest_Day_trailer.jpg *License*: unknown *Contributors*: trailer screenshot (20th Century Fox)

Image:LHMainTitle.jpg *Source*: http://en.wikipedia.org/w/index.php?title=File:LHMainTitle.jpg *License*: unknown *Contributors*: Florka, Iam4Lost, Kathleen.wright5, MegX, Sfan00 IMG, Stan Shebs

File:Little House on the Prairie Melissa Gilbert 1975 Crop 1.jpg *Source*: http://en.wikipedia.org/w/index.php?title=File:Little_House_on_the_Prairie_Melissa_Gilbert_1975_Crop_1.jpg *License*: unknown *Contributors*: NBC Television Network

File:Comedy Central 2011 Logo.svg *Source*: http://en.wikipedia.org/w/index.php?title=File:Comedy_Central_2011_Logo.svg *License*: unknown *Contributors*: MTV Networks (Viacom).

File:Comedy Central logo (1990s).svg *Source*: http://en.wikipedia.org/w/index.php?title=File:Comedy_Central_logo_(1990s).svg *License*: unknown *Contributors*: Svgalbertian

File:SouthParkBoys.jpg *Source*: http://en.wikipedia.org/w/index.php?title=File:SouthParkBoys.jpg *License*: unknown *Contributors*: Asenine, Fortdj33, Gurch, Jj98, Kurtis, Landon1980, Langston Bonasera, Mister Alcohol, Rodhullandemu, Skier Dude, The Twenty Thousand Tonne Bomb, TravisTX, 6 anonymous edits

File:Comedy Central logo.svg *Source*: http://en.wikipedia.org/w/index.php?title=File:Comedy_Central_logo.svg *License*: unknown *Contributors*: Adrianwo, Alex43223, Bigman17, Bongwarrior, ElSaxo, PudNoob

File:Comedy Central 2011 HD.svg *Source*: http://en.wikipedia.org/w/index.php?title=File:Comedy_Central_2011_HD.svg *License*: unknown *Contributors*: MTV Networks (Viacom).

Image:Knots Landing opening.jpg *Source*: http://en.wikipedia.org/w/index.php?title=File:Knots_Landing_opening.jpg *License*: unknown *Contributors*: CBS

Image:knotslanding1.jpg *Source*: http://en.wikipedia.org/w/index.php?title=File:Knotslanding1.jpg *License*: unknown *Contributors*: Juppiter, Sfan00 IMG, TAnthony, Thunderbrand, 1 anonymous edits

File:Flag of the United States.svg *Source*: http://en.wikipedia.org/w/index.php?title=File:Flag_of_the_United_States.svg *License*: unknown *Contributors*: Anomie

File:Flag of Florida.svg *Source*: http://en.wikipedia.org/w/index.php?title=File:Flag_of_Florida.svg *License*: unknown *Contributors*: Awg1010, Dbenbenn, Denelson83, DenghiùComm, Dzordzm, Ebyabe, Fry1989, Himasaram, Homo lupus, Hystrix, Madden, Mattes, Nightstallion, Pumbaa80, Reach Out to the Truth, Reuvenk, Rocket000, Vantey, Zscout370, 13 anonymous edits

File:Movita in Mutiny on the Bounty trailer.jpg *Source*: http://en.wikipedia.org/w/index.php?title=File:Movita_in_Mutiny_on_the_Bounty_trailer.jpg *License*: unknown *Contributors*: Trailer screenshot

MIX
Papier aus verantwortungsvollen Quellen
Paper from responsible sources
FSC® C105338

Printed by Books on Demand GmbH, Norderstedt / Germany